Inclusive School D

Acknowledgements

This guide was produced for the DfEE

by:

SENJIT (Special Educational Needs Joint Initiative for Training)
at University of London, Institute of Education

with:

Cullinan And Buck Architects Ltd.

DfEE would like to thank the following authors:

Mary Hrekow
Helen Clark
Flora Gathorne-Hardy
David Hayhow Associates

DfEE would also like to thank the following members
of the Steering Group for their contributions:

Cathie Bull	HMI, OFSTED
Mike Diaper	SEN Division, DfEE
Colin Jefferson	Property Officer, Cumbria LEA
Nick Peacey	Equal Opportunity Manager, QCA
Caroline Roaf	NASEN
Philippa Russell	National Children's Bureau
Robin Thomas	SEN Manager, Hampshire LEA
Peter Weston	Assistant Director of Education, East Sussex LEA

DfEE Project Team

Mukund Patel	Head of Schools Building and Design Unit
Beech Williamson	Principal Architect
Tamasin Dale	Research Architect
Mike Bubb	Editor

Published with the permission of DfEE on behalf of the Controller of Her
Majesty's Stationery Office.

First published 2001

ISBN 0-11-271109-X

Printed in the United Kingdom for The Stationery Office

Inclusive School Design

Accommodating pupils with special educational needs and disabilities in mainstream schools

Building Bulletin 94

Schools Building and Design Unit

Department for Education and Employment

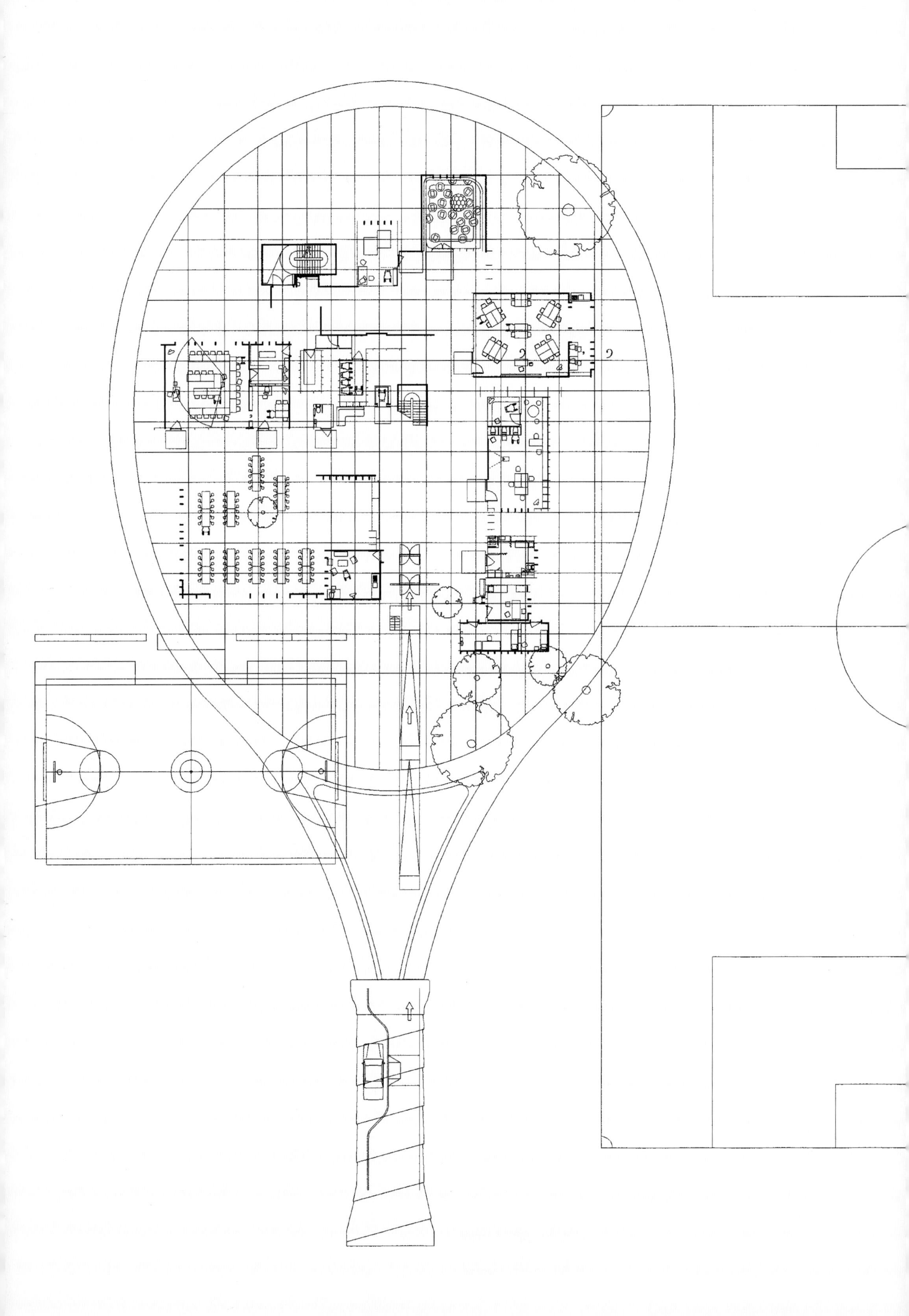

The principles of inclusive school design

The inclusive school.

The drawings in this book show some ways in which inclusive design can be integrated into the school and how this integration may be interpreted.

They do not show an entire school or prescribe the extent or disposition of inclusive provision.

Their purpose is to provoke all manner of thinking relating to inclusive design.

Inclusive design tries to break down unnecessary barriers and exclusions. In doing so, it will often achieve surprising and superior design solutions that benefit everyone.

The *National Curriculum* sets out three main goals for inclusive educational practices: to set suitable learning challenges; to respond to pupils' diverse learning needs; and to overcome potential barriers to learning and assessment for individuals and groups of pupils.

These educational goals are reflected in, and supported by, the following principles which relate to the design of new buildings or adaptation of existing accommodation and which inform this guide.

- Alongside other members of the school community, pupils with *special educational needs* and *disabilities* should expect to be treated with dignity and respect as individuals, with their needs being accounted for in the school surroundings.

- The school should provide an environment that is welcoming, safe and suitable for the educational needs of all pupils, including those with *special educational needs* and *disabilities*.

- The school buildings and grounds should be designed to enable all members of the school community to enter and move around the building so they can enjoy and participate in all aspects of school life to the best of their abilities and interests.

This guide does not present an ideal architectural blueprint for an inclusive school. A number of variables make it impossible to prescribe particular design solutions, including differences in relation to the location of each school, the design of existing buildings, the nature of pupils' needs and the particular policies and priorities of those involved in the planning for educational provision. The built environment, however well designed, cannot address all of the issues that might be experienced by pupils with *special educational needs* and *disabilities*. Managerial or organisational decisions beyond the scope of this guide often determine whether a pupil feels included or excluded.

Introduction

This Bulletin provides advice and guidance on how to accommodate pupils with *special educational needs* and *disabilities* in *mainstream schools*.

It will be of interest to governors and managers of primary and secondary schools, architects and landscape architects, other building professionals, local education authorities and other education providers as well the wider community of people who use school buildings and grounds.

Alongside Building Bulletin 91, 'Access for Disabled People to School Buildings' the illustrated text examines issues of physical access. It also addresses how to meet the design needs of pupils with sensory, learning, and *emotional and behavioural difficulties*. Practical recommendations are included with case studies that demonstrate how becoming more inclusive can bring benefits to the lives of all pupils who study, rest and play in *mainstream schools*.

PART 1 Explores why design matters and how it can affect the lives of pupils with *special educational needs* and *disabilities* in *mainstream schools*. The aim is to raise issues and pose questions that are dealt with in more detail in the sections that follow.

PART 2 Looks at how schools are preparing for, and accommodating, the physical design needs of pupils with *special educational needs* and *disabilities*. The case studies look at the issue from various perspectives and scales.

PART 3 Examines different forms of practical provision from whole school issues to the detail of furnishings, equipment and auxiliary aids. This supplements other forms of non-statutory guidance provided by the DfEE on school design.

PART 4 Contains further information, including references and useful contacts, and how the guide was produced.

Contents

PART 3 Provision

PART 4 Useful Information

Using this guide

AIM

All terms in italics are defined within the glossary at the end of this guide.

How can everyone working with mainstream schools ensure that the physical design of buildings, grounds, furniture and equipment enrich rather than constrain the lives of pupils with *special educational needs* and *disabilities*? The overall aim of this guide is to help people address this question strategically and work to create inclusive learning environments that benefit all members of the school community.

AUDIENCE

This guide forms part of a range of guidance produced by the Department for Education and Employment (DfEE). It offers advice to governors and managers of primary and secondary schools, architects and landscape architects, other building professionals, local education authorities and other education providers. It is also hoped that teaching staff, parents, pupils and members of the local community can find useful information within its pages.

SCOPE

Part 4.1 contains a list of relevant DfEE publications.

This guide raises issues, poses questions, stimulates discussion and provides a palette of technical specifications that can be drawn upon as part of the on-going process of creating more inclusive schools. The ideas presented can be used to help strategic planning for new provision, strengthen the development of architectural briefs, and guide day-to-day decisions about the maintenance and adaptation of existing buildings and grounds.

The phrase used throughout this guide is 'becoming more inclusive'.

The drawing on page 6 is compiled by gathering together some of the school spaces, furniture and details from other guidance and from the case studies found in this guide, to form a picture of an imagined inclusive school. Versions of this drawing appear at the beginning of parts 1, 2 and 3. Whilst indicating an effective range of minimal provision, their aim is to illuminate both the scope and the detailed conditions of inclusive design. By not being exhaustive they avoid prescribing fixed and limiting solutions. Sections of this drawing are shown at different scales throughout the guide revealing some of the possible relationships between a school's parts that can aid and promote inclusion and accessibility.

The child's tennis racquet, which is a design adapted from an adult tennis racquet, is shown throughout the guide in different ways as a reminder that inclusion is made possible by the whole school, the details found within it and the way it is used.

Navigation

Colour identifies which part of the book you are in.

Each part begins with a summary of its contents.

Left hand shoulder quotes lead the reader to other parts of the guide and publications or contacts outside of the guide.

PART 1 Design matters

Wayfinding, Comfort and Environment.
Key.
Red - different parts of the school need special surfaces for orientation and utility.
Yellow - clear circulation routes.
Mauve - a variety of external spaces.
Arrows - signage at key circulation points.

Inclusion considers school life from the perspective of the pupils and seeks ways to provide them with choice, dignity and self-esteem through the design of inclusive learning environments. This first section raises issues and asks questions about how careful changes to the physical design and management of school premises can help realise this goal. The issues considered are important for both new build schemes and adaptations to existing schools.

1.1 CONTEXT

Those working with schools will be aware of proposed legislation on SEN and Disability Bill, which will make it unlawful for education providers to discriminate against a disabled child. Further information can be obtained from the Disability Rights Commission's Advice Line. Contact details are included in Part 4.3.

The Government has committed itself to actively promoting the inclusion of pupils with *special educational needs* and *disabilities* within *mainstream schools* as part of a wider range of policies that recognise and celebrate human diversity. Within this policy framework, local education authorities play a strategic planning role, working in partnership with special and *mainstream schools* to realise this principle of inclusive educational practice.

These different arrangements by no means comprise a definitive list of the wide range of partnerships that currently exist.

There are many different approaches being adopted. In some cases, partnerships exist between special and/or mainstream schools located on different sites. In this situation, pupils and/or staff travel between the two schools sharing facilities and skills. *Co-location* is another option, where two schools are linked on the same site but retain their distinct identities. Here, there may be more of a physical overlap with pupils sharing spaces, such as assembly halls, sports facilities, as well as teaching resources. And, as stated above, all *mainstream schools* are required to consider resource provision for current and future pupils with *special educational needs* and *disabilities*.

The Centre for Studies on Inclusive Education has produced an index to help schools take a long term approach towards developing inclusive educational practices. Part 4.3 provides contact details.

15

Captions to drawings and photographs appear in the left hand margin.

Walking down a ramp.

BENEFITS FOR THE WHOLE SCHOOL COMMUNITY
As with teaching, asking questions about the physical design needs of pupils with *special educational needs* and *disabilities*, is not separate from, but an extension of a process of understanding how to enhance the lives of all pupils who study, rest and play in *mainstream schools*. It is difficult to overstate the importance of this point.

Case studies are coloured to match the part of the guide in which they are situated. They are included as examples that illustrate the breadth and detail of the subject of inclusive design.

ATTITUDES TOWARDS INCLUSION

Schools visited that have already established inclusive practices identify many benefits for the whole school community, including:

- a shift in attitude towards pupils with *special educational needs* and *disabilities* that encourage those involved to take a positive approach to meeting the physical design needs of all pupils;
- positive changes to pupil behaviour arising from improvements to the layout and internal design of shared areas such as corridors, cloak rooms and dining halls;
- improvements to acoustic conditions that help pupils with *hearing impairments*, as well as those

17

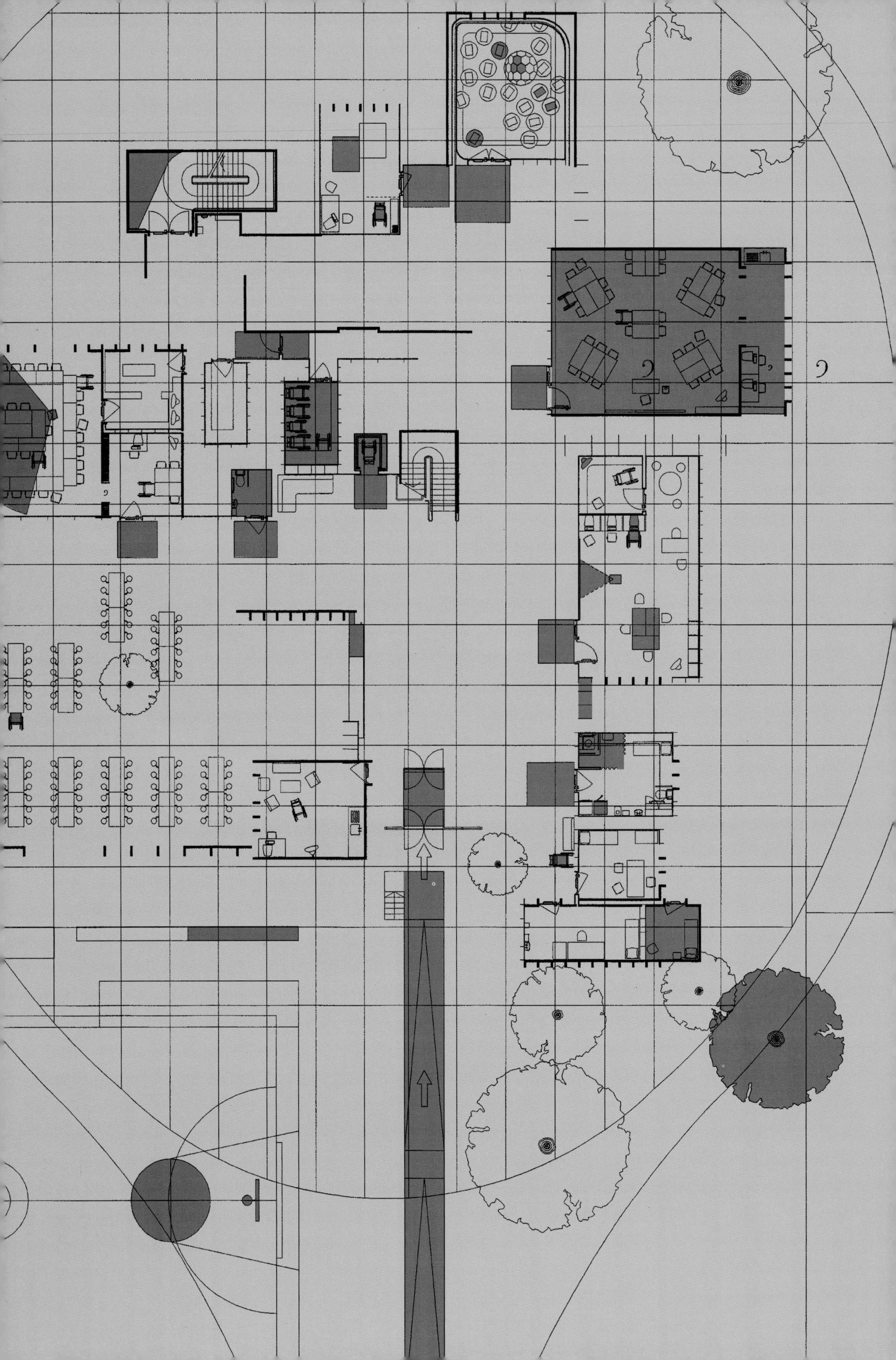

PART 1 Design matters

Special Needs.

Inclusion considers school life from the perspective of the pupils and seeks ways to provide them with choice, dignity and self-esteem through the design of inclusive learning environments. This first section raises issues and asks questions about how careful changes to the physical design and management of school premises can help realise this goal. The issues considered are important for both new build schemes and adaptations to existing schools.

1.1 CONTEXT

Those working with schools will be aware of proposed legislation on SEN and Disability Bill, which will make it unlawful for education providers to discriminate against a disabled child. Further information can be obtained from the Disability Discrimination Act Helpline. Contact details are included in Part 4.2.

The Government has committed itself to actively promoting the inclusion of pupils with *special educational needs* and *disabilities* within *mainstream schools* as part of a wider range of policies that recognise and celebrate human diversity. Within this policy framework, local education authorities play a strategic planning role, working in partnership with special and *mainstream schools* to realise this principle of inclusive educational practice.

These different arrangements by no means comprise a definitive list of the wide range of partnerships that currently exist.

The Centre for Studies on Inclusive Education has produced an Index to help schools take a long term approach towards developing inclusive educational practices. Part 4.2 provides contact details.

There are many different approaches being adopted. In some cases, partnerships exist between special and/or mainstream schools located on different sites. In this situation, pupils and/or staff travel between the two schools sharing facilities and skills. *Co-location* is another option, where two schools are linked on the same site but retain their distinct identities. Here, there may be more of a physical overlap with pupils sharing spaces, such as assembly halls, sports facilities, as well as teaching resources. And, as stated above, all *mainstream schools* are required to consider resource provision for current and future pupils with *special educational needs* and *disabilities*.

The DfEE and the Royal Institute of British Architects (RIBA) School Client Forum have collaborated in the publication of 'A Guide for School Governors: developing school buildings'. This document provides an overview of the design process and is obtainable from the DfEE and the RIBA.

Unless someone has had a technical training in, say, architecture or landscape design, the process of making decisions about the design of school premises may seem daunting. Yet, we all know that some places seem confusing and uncomfortable while others might make us feel at ease and able to concentrate or relax.

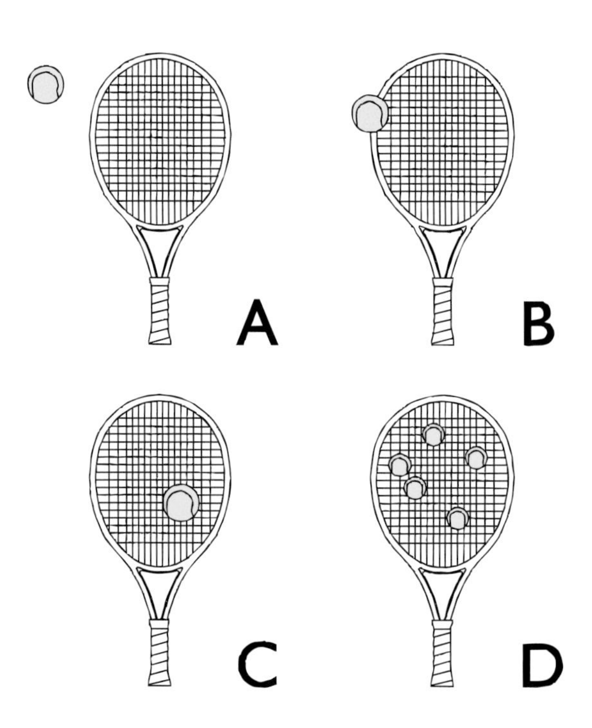

Moving towards inclusion. There are many ways in which special and mainstream schools are working towards partnership. A: special and mainstream are distinct and held apart. B: special and mainstream remain distinct but are housed in the same building with separate access. C: special and mainstream are distinct, housed in the same building with shared access. D: special and mainstream are completely integrated.

The challenge is to build upon such personal knowledge and develop our understanding of pupils' different bodily perceptions, experiences of movement and interaction with the architecture of school buildings and grounds. Does the entrance area feel welcoming? Do the grounds provide areas for quiet reflection? Are the classrooms accessible? Asking such questions helps identify how design changes - subtle adjustments to colours, lowering shelves or relocating libraries - can enable pupils with *special educational needs* and *disabilities* to participate to the full in the life of a *mainstream school*.

The table set out in Part 1.4 provides a framework within which such questions can be explored.

1.2 BECOMING MORE INCLUSIVE

As a school becomes more inclusive, there are several general considerations that need to be borne in mind.

Walking down a ramp.

BENEFITS FOR THE WHOLE SCHOOL COMMUNITY

As with teaching, asking questions about the physical design needs of pupils with *special educational needs* and *disabilities*, is not separate from, but an extension of a process of understanding how to enhance the lives of all pupils who study, rest and play in *mainstream schools*. It is difficult to overstate the importance of this point.

ATTITUDES TOWARDS INCLUSION

Schools visited that have already established inclusive practices identify many benefits for the whole school community, including:

- a shift in attitude towards pupils with *special educational needs* and *disabilities* that encourage those involved to take a positive approach to meeting the physical design needs of all pupils;
- positive changes to pupil behaviour arising from improvements to the layout and internal design of shared areas such as corridors, cloak rooms and dining halls;
- improvements to acoustic conditions that help pupils with *hearing impairments*, as well as those

with temporary hearing difficulties caused by ear infections. All pupils are better able to hear adults and each other;
- enrichment of the school grounds through the creation of better boundaries between different activities and a broader range of spaces for pupils to enjoy;
- the creation or adaptation of existing spaces to provide extra facilities such as small group rooms, storage space and laundries.

DESIGN AS A PROCESS

Becoming more inclusive is a process and not an event. It is not possible to present a checklist of design changes which, once completed, will ensure that a new school or adapted building is inclusive. Instead, inclusion is an on-going process of evaluating and adjusting the physical environment of school buildings and grounds. Such adjustments can be integrated into existing processes of decision making about the design and management of schools, which include:

Further information on 'Asset Management Plans' can be obtained from the DfEE website www.dfee.gov.uk/amps/index.htm

- changes that might be made in response to an individual child's particular needs;
- everyday management and maintenance programmes as set out within the school's Asset Management Plan;
- strategic planned improvements carried out in partnership with local education authorities' new build work.

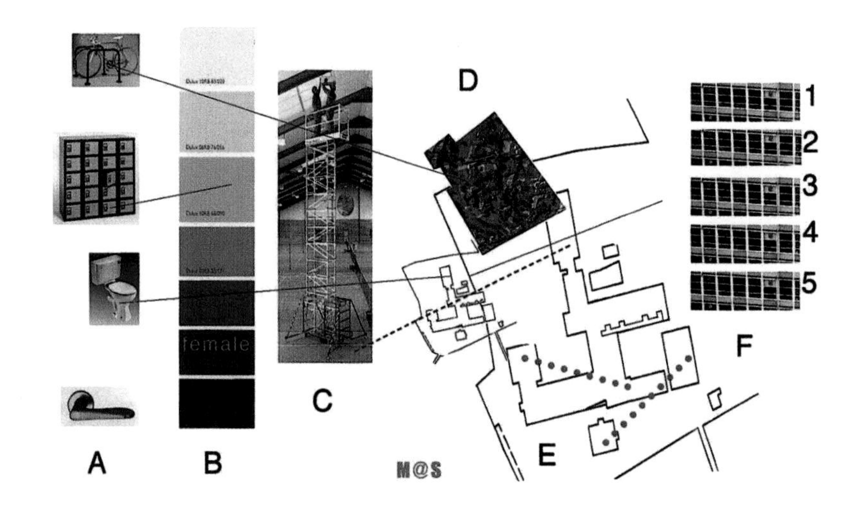

Integrating the principle of inclusion into different levels of decision making. A: small scale localised improvements. B: co-ordinated colour schemes, floor finishes, lighting etc. C: small building additions and removals. D: land use, audit of school grounds. E: space management. F: redevelopment.

DIVERSE AND CHANGING NEEDS

The design needs of all pupils are diverse and changing. Physically, emotionally and intellectually, pupils respond in different ways to different learning environments. Some flourish on the sports field while others prefer more intimate places to talk. Some are excited by the therapeutic value of water while others feel nervous. Some are drawn to practical workshops while others gravitate towards the computer screen. Pupils with *special educational needs* and *disabilities* are part of this diversity of abilities, interests and ambitions.

The 'Schools Special Educational Needs Policies Pack' (1995) published by the National Children's Bureau contains useful information on the diversity of pupils abilities and needs.

Children of the same age can be different sizes.

A pupil's physical design needs also change as they grow and mature. Some pupils may require additional space and facilities in order to manage their own personal care. For instance, a pupil with an indwelling catheter may initially need sufficient privacy and space for an assistant to provide practical help and advice, but as the pupil becomes more independent and confident, he/she may need accessible accommodation with good access to storage, waste disposal and washing facilitates in order to manage his/her own care.

There are a number of publications that explore the process of identifying and assessing the needs of pupils with special educational needs and disabilities, including: 'Implementing Effective Practice' (1998) by Janet Tod, Frances Castle and Mike Blamires, published by David Fulton; 'The SENCO Guide' (1997) DfEE; and 'Implementing the Code of Practice: Individual Education Plans' (1995) by Stella Warin, The National Association of Special Educational Needs.

The principles of maximising choice and control underscore what is described as the independent living movement. The Disabled Living Foundation helps people put these principles into practice by providing advice on use of equipment. Part 4.2 provides contact details.

Over longer periods of time, the design needs of a pupil with a physical disability may change as the nature of his/her disability changes, or because he/she has undergone medical treatment or experienced periods of illness. Again, there is not necessarily anything special about needing adjustments to furniture or the way a room is arranged.

POTENTIALLY CONFLICTING NEEDS

There may be potential conflicts between the different physical design needs of pupils. A pupil with limited mobility may require a room temperature that is uncomfortable for more physically active pupils. A pupil with *autistic spectrum disorder* or immature social skills may feel distressed in a large teaching space where other pupils are enjoying noisy interaction. Those involved will need to anticipate and plan to minimise such conflicts. This can be achieved by developing design solutions that allow for choice and control, such as providing shallow steps as well as ramps or including window shades as well as good natural light.

This issue of potentially conflicting needs is examined in greater detail in Part 1.3.

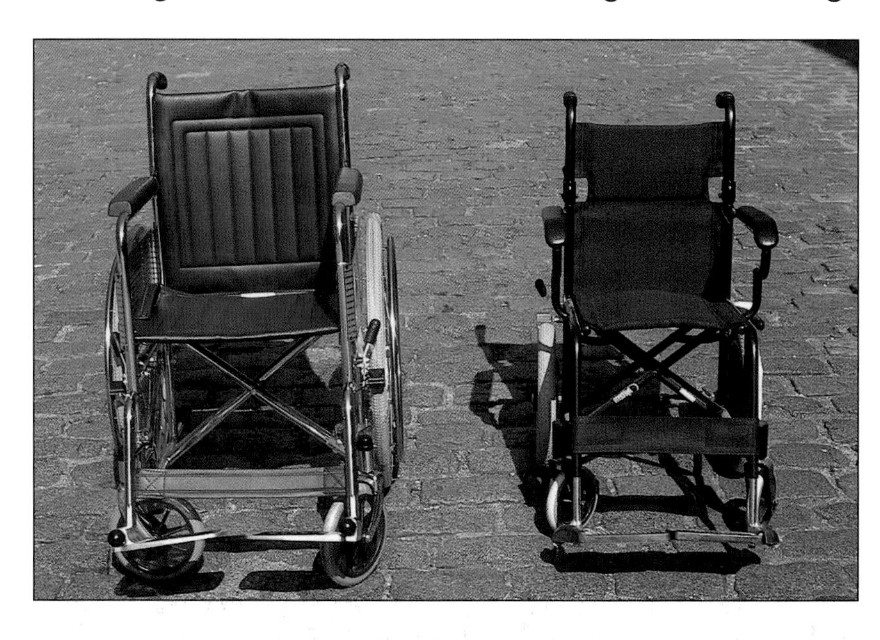

Children of the same age can be different sizes.

CREATIVE PROBLEM SOLVING

There are also resource constraints placed upon those working with schools that mean an ideal situation may not always be easily achieved. Creative problem solving is required. For example, a pupil with partial sight might require high levels of natural lighting that cannot be gained without significant structural changes to the

The Council for Disabled Children provides information and support for children with special educational needs and disabilities as well as their families and carers. Part 4.2 provides contact details.

Information about government funding for becoming more inclusive can be obtained directly from the DfEE. The Centre for Studies on Inclusive Education also provides useful sheets. Part 4.2 provides contact details.

room. Instead, the use of appropriate light bulbs and task specific light fixtures might provide an optimum (rather than ideal) solution.

PLANNING AHEAD

Thinking about and solving such problems takes time, as does the evaluation necessary to ensure that the pupil's needs continue to be met. This means that those working with, and in, schools must develop policies and practices that allow them to plan ahead, make changes in time and structure the on-going process of evaluation. As noted above, anyone working with schools must think about *special educational needs* and *disabilities* throughout the planning process.

PLANNING AHEAD

Whitmore High School in Harrow balances the task of responding to the particular needs of individual pupils with a longer-term approach that seeks to anticipate future needs. As an illustration of this approach, the decision was made to include floor drains during the adaptation of a toilet. Although there was no current need for a shower, one was likely to be needed in the future and installing drains in advance was far more cost effective than fitting suitable plumbing at a later date.

A mounted drill showing how simple changes can assist pupils who have difficulty using conventional hand tools. Whitmore High School, Harrow.

COLLABORATION

Finally, collaboration is needed between different people at different stages of the planning process, and to varying degrees. The pupils themselves need to be involved and able to express their views and aspirations. Parents have an important role to play in identifying needs and solutions, while local authority personnel and voluntary organisations can provide guidance, information and resources. Other people such as design professionals, therapists, and teachers can all bring their expertise to this process of creative problem solving.

Checklist of people who may need to be consulted in planning school design.

CHECKLIST

Architects and other design professionals.
Education and health specialists.
Governors.
Local authority representatives.
Parents/Carers.
Pupils.
Non-teaching school staff.
School teaching staff.
Therapists.
Voluntary organisations and other agencies.

1.3 BASIC QUESTIONS

Part 3 offers practical advice on how particular forms of provision can help address these questions.

The task is now to look at these design issues in more detail and ask some basic questions about the following design criteria. The aim is to help raise issues and prompt discussion about the implications of becoming more inclusive.

See Part 3.2 onwards for practical recommendations regarding the provision of adequate space.

More information about these space requirements can be found in Building Bulletin 77 'Designing for Pupils with Special Educational Needs: Special Schools' as well as the more recent Building Bulletin 91 'Access for Disabled Pupils to School Buildings'.

ADEQUATE SPACE AND PROVISION OF SPACES

- What space is needed?
- Is it possible to modify existing space to accommodate pupils with *special educational needs* and *disabilities* within the school building and grounds?
- How can areas be used flexibly or new places created in order to provide the specialist spaces needed, such as rooms for physiotherapy or counselling, medical rooms, storage for equipment and so forth?
- Can the organisation of the school be adjusted to relieve any pressure on space?

Anyone who has travelled in a train at rush hour knows the strain of being in a crowded space: obstacles barring movement; distracting noise levels; physical discomfort and mental stress. The same problems are faced by pupils and adults working in

Part 3 provides a list of the kinds of additional spaces that a mainstream school may need to develop in becoming more inclusive.

Building Bulletin 77 'Designing for Pupils with Special Educational Needs: Special Schools' includes detailed information on these additional areas.

Building Bulletin 91 'Access for Disabled People to School Buildings' provides information on issues of physical access.

Part 3.1 provides detailed guidance on improving access within mainstream schools. This guidance builds upon that set out in Building Bulletin 91 'Access for Disabled Pupils to School Buildings'.

areas where there is inadequate space for a group and its needs.

Becoming more inclusive involves ensuring that there is adequate space to accommodate pupils with *special educational needs* and *disabilities* within the general areas of school buildings and grounds. So, for instance, it is important to address the need for space for circulation and for pupils who use wheelchairs or sticks, both in corridors and teaching spaces. A pupil with a physical *disability* may feel distressed in a busy playground. Creating quieter play areas and extra seating may help solve the problem.

As well as finding creative ways to flexibly use and manage general school areas, it is also important to consider the provision of specialist places, such as therapy rooms, storage areas for communication aids and fully accessible toilets. Issues relating to manual handling and the health and safety needs of staff might also have space implications, both for general areas and specialist places within the school.

The Government has recognised all these space implications of becoming more inclusive and revised its methods of assessing school capacity in three main ways. Allowances will be made for:

- additional areas in individual rooms or spaces to allow appropriate access and use, including classrooms (where an area at the top of the Area Guidelines range would be needed);
- specialist support spaces for pupils with *special educational needs* and *disabilities* on the register of mainstream classes;
- specialist classbases, which may be used for the majority of the school day by those with specific or severe difficulties, such as severe *learning difficulties*, who are on the register of mainstream classes.

PHYSICAL MOVEMENT

Can pupils with mobility impairments get safely around the school?

Pupils with mobility difficulties, or who have difficulty walking long distances are likely to face physical barriers such as steps, narrow doorways and long corridors. What appear to be small details can have large consequences: even a seemingly low threshold strip can be a barrier. Pupils may also face difficulties leaving the school safely if there is an emergency such as a fire. Other examples of such physical barriers include slippery outdoor surfaces, heavy doors and unsuitable door handles.

WAYFINDING

Can pupils orientate themselves and find their way around the school buildings and grounds?

A scented pathway.

Wayfinding looks beyond physical barriers and includes a broader set of design issues that affect how people orientate themselves and avoid getting lost. The process of planning and making journeys is affected by personal factors, such as how pupils perceive the built environment and their ability to orient themselves spatially and to process information. As one example, pupils with certain forms of *autistic spectrum disorder* can suffer increased levels of anxiety if the building is difficult to understand, which can lead to stress and challenging behaviour in the classroom. Whether they are familiar with the area and what information they have been given about it will also affect their journey. Some pupils may not use the written word. For them, the use of colours, textures and symbols within wayfinding systems become especially important.

The NHS Estates report, 'Wayfinding', includes a useful introduction to these issues. Organisations such as Mencap, who work with people with learning difficulties are also able to offer advice. Part 4.2 provides contact details.

Part 3.1 provides practical guidance on wayfinding.

The Royal National Institute for the Blind (RNIB) and the Joint Mobility Unit have collaborated in the publication of 'The Sign Design Guide'. Part 4.2 provides contact details.

Section 3.1 provides practical guidance on lighting.

Features of the environment play an important role in helping pupils find their way. Examples include:

- the design and location of signs;
- how easy it is to identify differences between areas in terms of their style, colour, size, and the noises and smells from activities taking place;
- the design of prominent landmarks for pupils to recognise;
- the complexity of the site, routes and interiors of buildings;
- the amount of visual clutter detracting from, or obscuring entrances, route ways, places of arrival and signs.

VISUAL ASPECTS

One of the most important ways people receive information about the built environment is through sight. Three main factors govern the way we receive visual information. These are the condition of the eye, the quality of the lightsource, and the nature of the object being viewed.

Are lighting levels, colours and other visual aspects designed to help pupils, especially those with *sensory impairments*, participate in school life?

What pupils with *visual impairments* can actually see will vary enormously. Some will see things clearly but within a very limited visual field, while others may have a loss of central vision. Other pupils may have a loss of acuity, a blurring of vision, or loss of colour vision. Many of these conditions can occur together. There are also pupils who are blind.

Fully sighted

General loss: Diabetic retinopathy

Central loss: Macular degeneration

Peripheral loss: Glaucoma

The RNIB produces a wide range of information on the experiences and design needs of people with visual impairments such as their publication 'Building Sight'.

The difficulties pupils with *visual impairments* experience and their responses to light will vary. The avoidance of glare from windows, roof lights or light fixtures is important for most pupils, but some will need additional illumination to carry out specific tasks or to ensure that teaching boards are clear. In some cases, the needs of different individuals will conflict. As one example, the use of higher than normal levels of lighting can help pupils whose visual *acuity* can be improved by the contraction of the iris, producing a greater depth of field. For others, such as pupils who need a dilated iris to see around a central opacity, these higher light levels cause problems.

Part 3.1 provides practical guidance on safe egress from buildings for pupils with sensory impairments.

The way rooms are arranged will impact upon the lives of pupils with *visual impairments*. Providing uncluttered route ways and thinking about large areas of glazing are important, as is the provision of mobility training. So, too, is making sure teaching resources and pupils' personal possessions are not moved unnecessarily.

Good colour contrasts help identify doors and door handles.

The DfEE publication 'Furniture and Equipment in Schools: a purchasing guide', offers useful advice as does the report 'Colour and Contrast' produced by the RNIB.

Colour and colour contrast is another important visual design consideration. Enhancing the colour and contrast of objects helps everyone under less than ideal lighting conditions, especially those with *visual impairments*, locate significant elements such as doors, door handles, changes in directions in corridors and changes in floor levels and steps. The colour of lighting, rooms and furnishings also affects pupils on a more subtle level. Green-tinged lighting creates a very different atmosphere from, say, red. Some wall colours seem to soothe pupils while others can be irritating or physically uncomfortable.

As discussed below, lighting is also important for pupils with *hearing impairments*. For instance, the teacher's face needs to be well lit to enable pupils to lip read more easily.

Part 3.1 provides practical guidance on acoustic design.

SOOTHING ENVIRONMENTS

Wigton Infant School in Cumbria has taken great care in the decoration of its internal spaces. As one example, most teaching spaces have an area within the room that is painted in pale colours and free of bright images. The teaching staff find this area provides a soothing place for pupils with certain forms of *autistic spectrum disorder* who become upset or overly distracted by too much visual stimulation. At the same time, the school recognises that other pupils benefit from colourful, stimulating environments.

ACOUSTIC ASPECTS

Building Bulletin 87 'Guidelines for Environmental Design for Schools' provides detailed information on these environmental issues.

Are acoustic conditions designed to help pupils, especially those with hearing impairments and other sensory impairments, participate in school life?

As with visual information, the sounds we receive are affected by three main factors: the condition of a person's auditory system; the source of sound; and aspects of physical design such as the size of the space and characteristics of the materials within it that affect acoustic performance.

There is huge variation in possible *hearing impairments* and the design requirements of pupils that arise. To ensure that pupils with *hearing impairments* can make

National Deaf Chidrens Society (NDCS) produces a wide range of publications on the experiences and design needs of people with hearing impairments.

maximum use of their residual hearing, it is essential to be aware of their ability to hear sounds of different frequencies. Some pupils will have greater residual high frequency hearing rather than low frequency. In manipulating acoustic conditions, it is necessary to consider the following.

- Noise levels.
- Reverberation time. This describes how long a sound takes to decay with clarity decreasing as the reverberation time increases.
- Acoustic absorption. The addition of absorbent surfaces reduces the reflected sound. This results in a reduction in the overall sound level and also in the reverberation time.
- Sound insulation for walls, floors and ceilings.

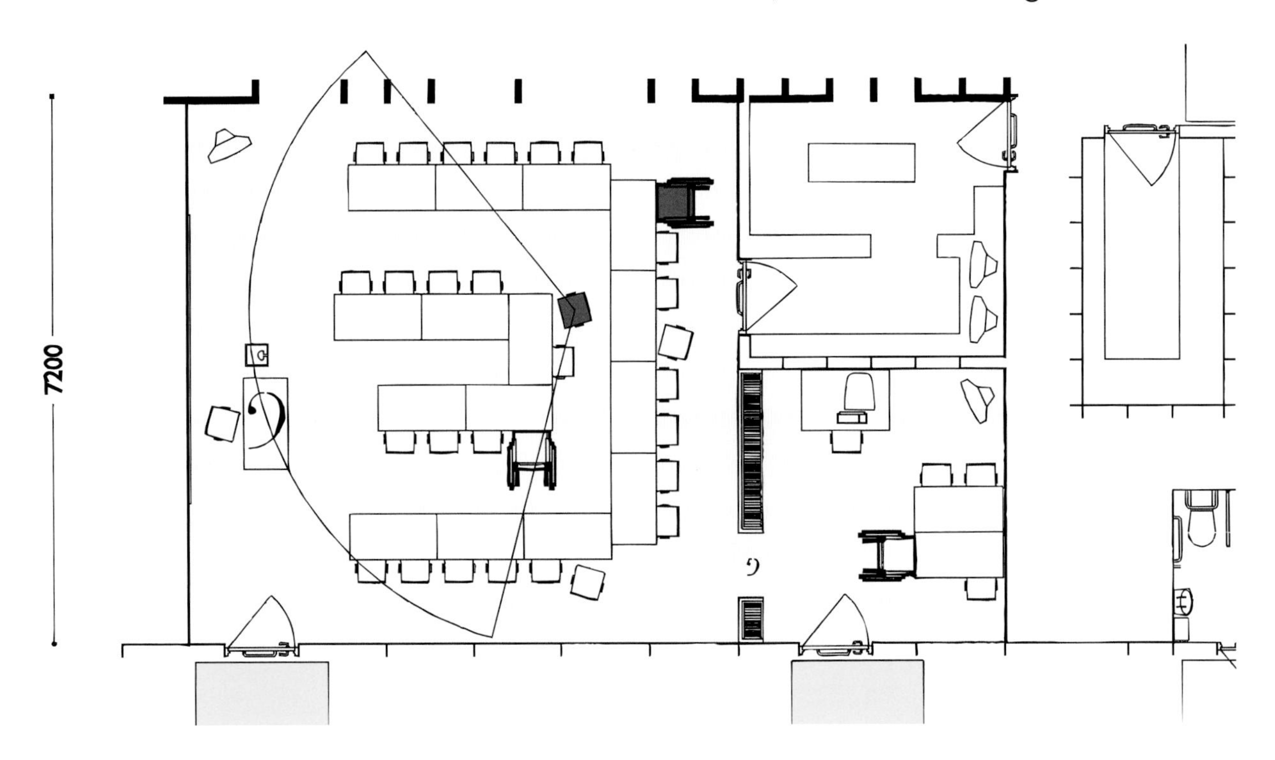

Classroom – Horseshoe Layout with Small Group Room and walk-in store adjacent. All pupils need good lighting and acoustics. Pupils with hearing impairment should be seated where they can have a good view of the teacher and still see other pupils' faces. Using the bass clef at different sizes, competing sound sources are shown in the drawing.

Noise can come from a number of different sources such as classroom activities, ventilation equipment and aircraft or road traffic.

The aim of good acoustic design is to enable people to hear clearly without distraction. Ways of manipulating acoustic conditions include:

- removing the source of the noise;

- reducing background noise levels, especially low frequency sounds;
- carefully planning where and when noisy activities take place;
- improving the sound insulation between spaces;
- reducing reverberation times by changing the acoustic absorption of teaching spaces;
- providing sound insulation between adjoining spaces by specifying adequate partitions, walls and floors;
- choosing appropriate materials, such as floor coverings. Some floor coverings muffle the sounds of footsteps, wheelchairs and other mobility aids and thus provide a quieter environment.

Acoustics is one example of the benefits that can accrue to the whole school community from strategic thinking about inclusive design.

Alongside addressing the needs of pupils with *hearing impairments*, there are other improvements to be gained from thinking about noise. One example is the benefit of creating quiet supervised spaces for pupils who at times find aspects of school life too intense or distracting. Here pupils can go to calm down, cry, reflect, work with a therapist or friend, or just play in private. As a second example, it is important to consider how pupils with partial sight or who are blind use background noises to help guide them around school premises. Noise should not be removed altogether, just controlled.

SENSORY ENVIRONMENTS AND TEMPERATURE

How can environments be designed so as to provide stimulating sensory experiences and comfortable temperatures for pupils?

Pupils enjoying a sensory room.

Sight, hearing, smell, touch, and taste are the five most familiar senses. A sixth sense, the haptic has been identified. This refers to the kinds of feelings you receive through the skin and frame of the body as you lie against a log, or sense water pushing against your hand.

Contact details for Learning through Landscapes are included in Part 4.2.

The recent DfEE publication 'Designing for 3 to 4 Year Olds' addresses the design needs of younger children.

There is a great deal of scope for thinking about how all these sensory experiences can be enhanced. Research carried out by the organisation Learning through Landscapes showed how all pupils benefit from school grounds that are furnished with scented plants, seating that is pleasant to the touch, and installations such as wind chimes that create interesting sounds. Certain groups of pupils, such as those with communication or *learning difficulties*, may find such stimulation and exploration of the senses especially pleasurable and important to their educational development. For all pupils, especially younger children, it is important that they are given the chance to explore the world through their haptic sense.

Part 3.1 provides practical guidance on temperature and ventilation.

The Disabled Living Foundation offers advice on furniture and equipment for people with disabilities. Part 4.2 provides contact details.

Temperature is another important consideration. As one example, pupils with limited mobility may not generate as much body heat as a fully mobile child and need higher room temperatures. Another pupil who is hyperactive may require relatively cool environments. Those working with schools will need to bear these potential issues in mind when maintaining heating systems, installing new systems and exploring ways of adjusting room temperature levels using things such as fans, blinds, and additional heating equipment.

Part 3.2 provides practical guidance on landscape design.

Ventilation also needs to be considered, especially within hygiene areas where there are potential problems of poor air quality.

FURNITURE AND EQUIPMENT

How can design details be improved and suitable furniture and equipment be provided to enable pupils to participate in school life?

Design details, such as the height of grab rails can have a very significant impact on pupils' lives. These need to be carefully checked and, where necessary, changed and adjusted to enable pupils to use facilities alongside their peers.

Part 3.1 provides practical guidance on furniture and equipment.

Part 3.2 provides practical guidance on toilets and personal care.

Part 3.1 provides practical guidance on electrical fittings and equipment.

The provision of suitable furniture and equipment is important and schools have tried many alternative strategies to solve problems. As one example, pupils with mobility difficulties may only be able to transfer themselves from their wheelchairs to the toilet if there are wall grab rails placed at the right height and distance from the toilet seat. For a child with *hearing impairments*, the provision of an appropriate *sound field amplification system* may be one solution to help them hear more clearly. Another pupil with moderate *learning difficulties* may require adjustments to computer key-boards to help him/her study independently. There are also portable pieces of equipment such as trolleys that can be used to move specialist equipment to different areas.

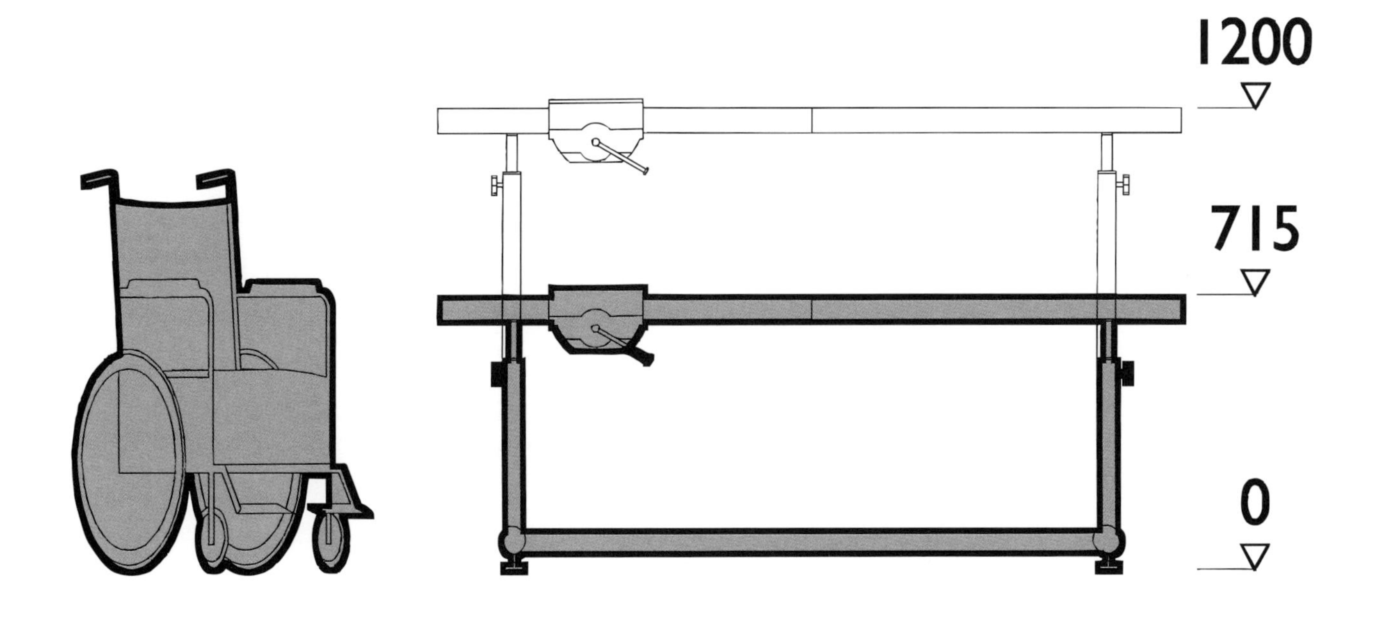

Adjustable work bench.

ADJUSTABLE FURNITURE

Treviglas Community College in Newquay accommodates pupils with a wide range of *special educational needs* and *disabilities*. One pupil who used a wheelchair to study science was unable to use some of the higher benches and work surfaces. The school had a chair custom made that is similar to an adjustable office chair with added foot rests that can rise higher to enable the pupil to use the benches. The chair can be easily transferred and used by other pupils.

1.4 PROMPTING DISCUSSION

CORE QUESTIONS, USING THE TABLE

The table on page 36 has been developed for governors, school staff and anyone else involved in planning for, and implementing design changes to mainstream schools.

The aim of the table is to prompt discussion. It takes stock of issues raised about the kinds of design changes that will need to be addressed to accommodate current or future pupils with *special educational needs* and *disabilities*. Each of the four headings in the left-hand column represents an area of learning and development, which pupils may experience difficulty with and which may have implications for one or more of the design aspects listed across the table. Examples of the range of *special educational needs* and *disabilities* likely to be affected are listed under each heading. Across the top of the table are the 'prompt questions'.

Once the initial discussion has taken place, governing bodies will need to look at some aspects in more detail and, drawing on Part 3, explore what kinds of practical provision are needed. They will also need to prioritise such changes within a strategic long-term development plan for the management, maintenance and building programme of buildings and grounds.

The table is also intended as a checklist that can be

photocopied and used for the planning of inclusive school provision. The imaginary example cited below helps to illustrate some of the points refered to.

OPPORTUNITY BASE

School A has been designated by their LEA to be an Opportunity Base for pupils with *special educational needs* and *disabilities* in the area of language and communication. The Opportunity Base has a fully inclusive policy and the children will be placed in mainstream classes and have access to the full mainstream curriculum with the support and advice of specialist teachers and trained learning support assistants.

However, it is clear from the profile of many of the pupils that the school will need a degree of specialist input in the area of language and communication. The school intends to identify small groups for intensive work in this area. The aim is also to include other pupils at early stages of the SEN Code of Practice within the school who would benefit from this intensive small group work.

There is no suitable teaching area that is free of noise and other distractions where the small group sessions could take place.

It is a large school which has had several linked extensions added over the years making it difficult to navigate. This has been an ongoing problem for all pupils and some adults. The school feels that this will be a particular issue for the pupils with language and communication problems who may have difficulty with general comprehension and be unable to ask for help with directions confidently. The Opportunity Base intends to introduce a sign and symbol system, to which all pupils in the school will be introduced.

Using the table, the school runs across the possible issues that might arise and identifies a need for additional space

As a result of this initial discussion, the school identifies two priority areas for development which are then included in the school development plan.

- Provision of a quiet, small group teaching space.
- Design of a wayfinding system, using symbols and key words to make it easier for all pupils, staff and visitors to find their way.

A series of actions are identified and target dates set for them to be completed. An under-used cloakroom space is then identified as suitable for conversion into a small group teaching room and the school now hopes to have completed this work within six months.

Moving across the table, the school also identifies an issue with wayfinding.

The provision of a wayfinding system is approached through a governing body contact with a higher education establishment in the form of a project undertaken by a small group of Art and Design Students. They are working with the specialist teacher in charge of the unit to develop ideas, which will be presented to the governing body for consideration.

CATALYST FOR DISCUSSION

- General spaces.
  Is it necessary to modify existing space or to reduce group sizes to accommodate pupils with *special educational needs* and *disabilities* and any adult assistant during classes and other activities, within the school building and grounds?

- Specialist spaces.
  How can areas be used flexibly, and what new rooms need to be created to provide the specialist spaces needed, such as rooms for physiotherapy or counselling, medical rooms, storage for equipment and so forth?

- Physical movement.
  Can all pupils with mobility difficulties get to, in, around and safely out of school?

- Wayfinding.
  Can pupils orient themselves and find their way around the school buildings and grounds?

- Visual aspects.
  Are lighting levels, colours and other visual aspects designed to help pupils, especially those with visual impairments, participate in school life?

- Acoustics.
  Are acoustic conditions designed to help pupils, especially those with hearing impairments and other sensory impairments, participate in school life?

- Sensory environments and temperature.
  Does the current environment provide stimulating sensory experiences and comfortable temperatures for all pupils?

- Furniture and equipment.
  Does any furniture need to be adjusted or suitable equipment provided to enable pupils to participate in school life?

Catalyst for discussion

	General space	Specialist spaces
COMMUNICATION AND INTERACTION		
Speech and Language Delay I Impairment I Disorder •	o	o
Specific Learning Difficulties I Dyslexia I Dyspraxia •	o	o
Autistic Spectrum Disorder •	o	o
Deafness I Hearing Impairment •	o	o
Deafblindness •	o	o
Visual Impairment •	o	o
COGNITION AND LEARNING		
Learning Difficulties •	o	o
Specific Learning Difficulties I Dyslexia I Dyspraxia •	o	o
Autistic Spectrum Disorder •	o	o
BEHAVIOURAL, EMOTIONAL AND SOCIAL DEVELOPMENT		
Emotional and Behavioural Difficulties •	o	o
Withdrawal I Isolation I School Phobia •	o	o
Disruptive and Disturbing Behaviour •	o	o
Hyperactivity and Lack of Concentration •	o	o
Challenging Behaviours •	o	o
SENSORY AND PHYSICAL DEVELOPMENT		
Deafness I Hearing Impairment •	o	o
Deafblindness •	o	o
Visual Impairment •	o	o
Physical Disability •	o	o

ACTION

Physical movement	Wayfinding	Visual aspects	Acoustics	Sensory environments	Furniture and equipment
o o o o o o	o o o o o o	o o o o o o	o o o o o o	o o o o o o	o o o o o o
o o o	o o o	o o o	o o o	o o o	o o o
o o o o o	o o o o o	o o o o o	o o o o o	o o o o o	o o o o o
o o o o	o o o o	o o o o	o o o o	o o o o	o o o o

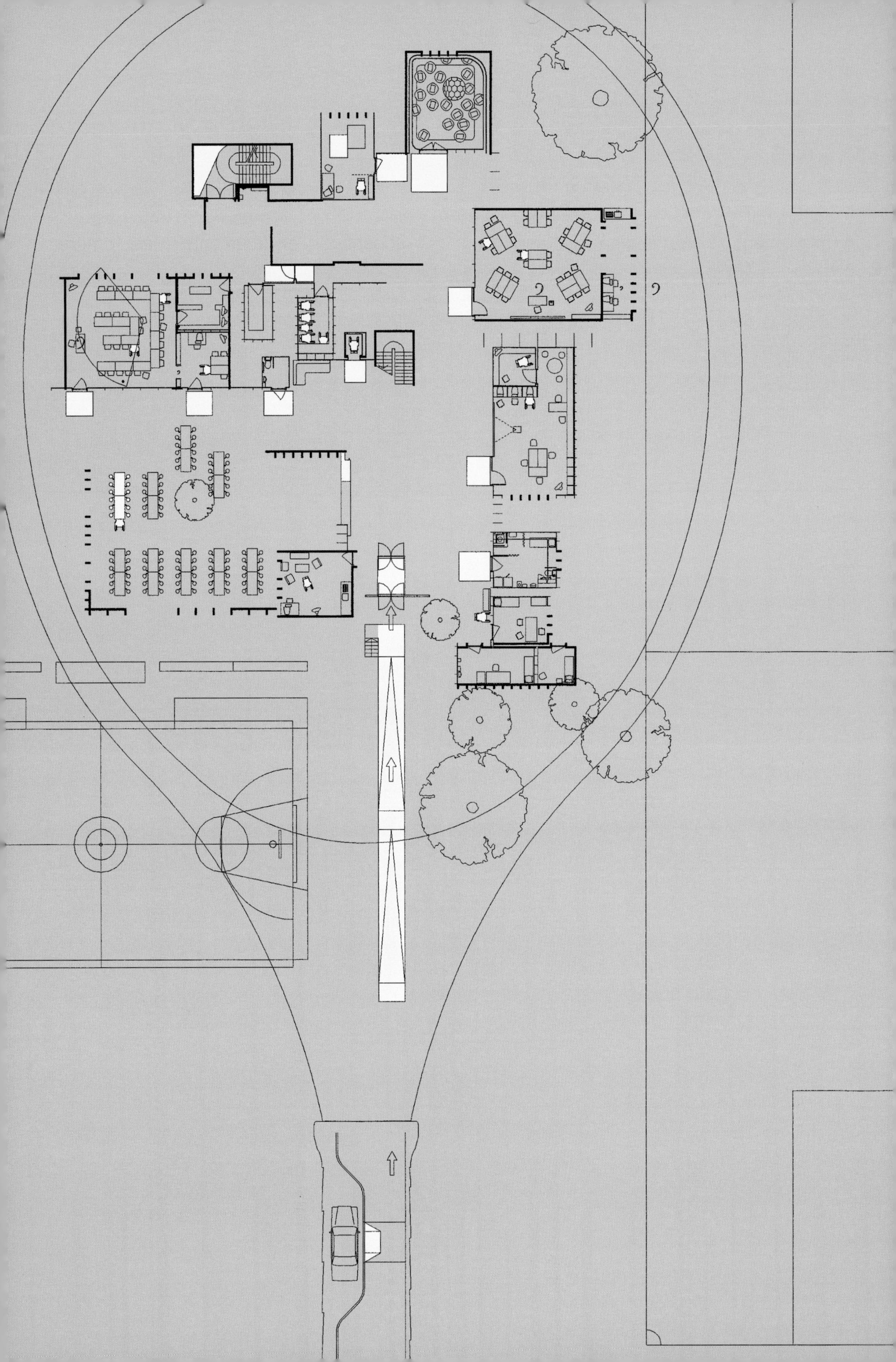

PART 2 Evaluation

Degrees of Inclusion.

This central part provides a bridge between the principles explored in Part 1 and the checklist of practical provision set out in Part 3. It reflects upon how schools are tackling issues of inclusive design in different ways, evaluating both examples of good practice and problems to be overcome. Each set of issues ends with a checklist that draws together case studies woven throughout the guide and offers a summary of issues to consider.

2.1 RESPONDING TO THE NEEDS OF A PUPIL

Many changes to the design of school buildings and grounds have come about in response to the needs of an individual pupil. Local Education Authorities are responsible for decisions about the allocation of resources. Increasingly, they are developing new partnerships to enable pupils to be matched with the right school and for the necessary design changes to be planned for and implemented well in advance of the pupil's first day at school. While it is right to use the opportunity of an individual joining the school to make changes, a broader perspective should be taken that adopts a long-term approach to becoming more inclusive.

LONG TERM PLANNING

Cornwall County Council has developed a long-term strategy for the intake and transfer of pupils with *disabilities* or those who have been identified as having *special educational needs*. This involves liaison between the health and education authorities, as well as close links with head teachers, parents, design professionals and other people involved in specialist educational provision.

For example, a pupil who had been paralysed from a neck injury was accommodated in a primary school close to her home. The Local Education Authority soon began to plan for her secondary education.

Checklist.
Adjustable furniture, room to pass, use of information technology, can be found in Part 1.

Having identified Bodmin Community College as a suitable partner, the process of planning for, and implementing, necessary design changes began two years in advance of her enrolment. Such changes involved the creation of a suitable parking bay, external ramps, internal lifts and a suitable toilet. Great care was taken to consult with the pupil too: for instance, measurements were taken of her wheelchair to make sure the dimensions of the toilet were appropriate. Thus, Bodmin Community College reviews its rolling programme of maintenance and building work in the light of anticipated design changes future pupils will require. It is also able to work with the County Council in identifying ways of meeting costs that arise once the pupil has enrolled.

A swimming pool with hoist.

SUMMARY

Reflecting upon the case studies, a number of core issues can be identified.

Design matters

For all pupils and, at times, especially for pupils with *special educational needs* and *disabilities* in *mainstream schools*, design affects their physical experiences, emotional lives and intellectual development.

Design is a means to an end

Design matters not as an end in itself but because it allows pupils to pursue other activities.

Design is a process

Creating inclusive learning environments is not an event but an on-going process of creative problem solving.

Long-term planning is crucial

In order to anticipate and accommodate the complex and changing needs of pupils within *mainstream schools*, time is needed to plan for, implement and evaluate the design needs of individual pupils.

2.2 BECOMING MORE INCLUSIVE

As well as responding to the particular needs of individual pupils, *mainstream schools* across the country are working to create more accessible and flexible learning environments that can be easily adapted as pupils' needs change.

ADAPTING A SCHOOL LAYOUT

The conventional model of a school layout is often based upon the assumption that pupils move through distinct areas of a school as they grow older. For many inclusive schools, the arrangement of where pupils of different ages study is challenged.

This is illustrated by Ullswater Community College in Cumbria. The school has over 1400 pupils, with approximately one fifth having *special educational needs*. There is a very wide range of abilities and needs, from pupils with severe learning difficulties to those seeking to attend university. The way the school is designed and spatially managed has to take into account this diversity of abilities and the fact that pupils need to move around the school to access different spaces, such as small group study rooms or rooms where one-to-one teaching can take place.

Orientation and wayfinding have become key issues, with the school examining both lighting and signage in circulation spaces. The fact that the school comprises many different buildings and is on a sloping site has presented difficulties, such as problems of physical access and the fact that pupils have to move long distances between teaching spaces.

The approach taken by the school is to be as creative as possible in identifying ways to redesign and fund changes to the school buildings

Example of an accessible toilet.

and grounds. Maintenance budgets are used to resurface floors and external areas, while short-term solutions include using paper blinds to reduce glare. Careful timetabling is used to tackle the problem of long journeys for pupils. Overall, the school recognises the on-going nature of becoming more inclusive.

EVOLVING FORMS OF PROVISION

As stressed in Part 1, there are many models for inclusive educational provision and different ways that schools physically link together. Ideas about good practice also continue to evolve within different contexts in response to changing needs and attitudes.

This process of continuing to refine policies and practices is illustrated by West St Leonard's Primary School in East Sussex. Originally, a *mainstream school* and a *special school* were combined within one building. At the time of building, the local authority felt that the schools should retain their distinct identities with a head teacher managing separate sets of new buildings. After a few years, however, the teaching staff, SEN coordinator, school governors and parents all agreed that it would be better for the two schools to combine and share all the buildings fully.

This decision has resulted in a number of benefits. For example, the whole school is able to use facilities originally designed for the *special school*, such as the laundry facilities, shower room and a suite of small rooms suitable for small group study. The medical room is well used by all members of the school.

Checklist.
- *Benefits for the whole school community*
- *Planning ahead*
- *Making room*
- *Soothing environments*
- *Managing noise levels*
- *The need for storage space*
- *Accessible library facilities*
- *Flexible nursery class base*
- *Parents' room*
- *Toilet and personal care provision*
- *Inclusive sports facilities*
- *Play spaces*

SUMMARY

Reflecting upon the case studies, a number of core issues can be identified.

An inclusive approach

Over and over, the importance of a creative and positive attitude towards inclusion is emphasised by those involved in the day to day management of schools. Inclusive design takes place when a commitment towards inclusive educational practices and equal opportunities permeates the culture of the school.

Developing practical solutions

Many examples of design changes can take place as part of on-going maintenance and building programmes that help schools become more inclusive. These include wayfinding, physical access, colour and contrast, and creative timetabling.

The need for adequate space

The lack of space for storing equipment and teaching resources is a concern common to most *mainstream schools* becoming more inclusive. Problems of over-crowding within classrooms also recur. Thinking about how to use space flexibly and create extra space are therefore core challenges for those involved in decisions about school design and management.

Changing approaches

The case studies provide a snapshot of a rapidly evolving field of policy and practice. Those working with schools have to be prepared to respond to fluctuations in pupil needs as well as new requirements that have an impact on how the school is designed and managed.

Benefits for the wider school community

The case studies confirm that thinking about inclusion and designing for pupils with *special educational needs* and *disabilities* will usually yield benefits for a much wider community of pupils, school staff, parents, and other children and adults seeking to access resources and courses provided on the school site.

2.3 LONG TERM PLANNING FOR SCHOOLS

Many of the case studies include examples of schools working in partnership with other agencies and local authority personnel to exchange ideas on inclusive design.

Checklist.
- *Long term planning adapting a school layout*
- *Evolving forms of provision improving governmental design guidance*
- *Designing a new school layout*

IMPROVING ENVIRONMENTAL DESIGN

Avon County Council has re-formed as four unitary authorities Bristol, South Gloucestershire, Bath and North-East Somerset, and North Somerset. A series of fora have been created to enable staff working with these unitary authorities to exchange ideas on environmental design. As one example, the head of the hearing impaired team has collated examples of acoustic problems and solutions shared by many schools. As a second example, support staff working with pupils with *visual impairments* meet regularly to reflect on their experiences and discuss general design recommendations for individual situations.

SUMMARY

Reflecting upon the case studies, a number of core issues can be identified.

Different contexts

Each school is working within a different context and in response to different needs. The links developed with other schools and education services also vary.

Sharing ideas and resources

Across this diversity, it is important for those working

with schools to seek ways to exchange ideas and develop long-term strategies that help plan for, and accommodate, the design needs of pupils with *special educational needs* and *disabilities* in *mainstream schools*.

Links
- Placement of a resource unit.
- Fire procedure.
- Creating quiet rooms.

Standard manual alphabet RNID.

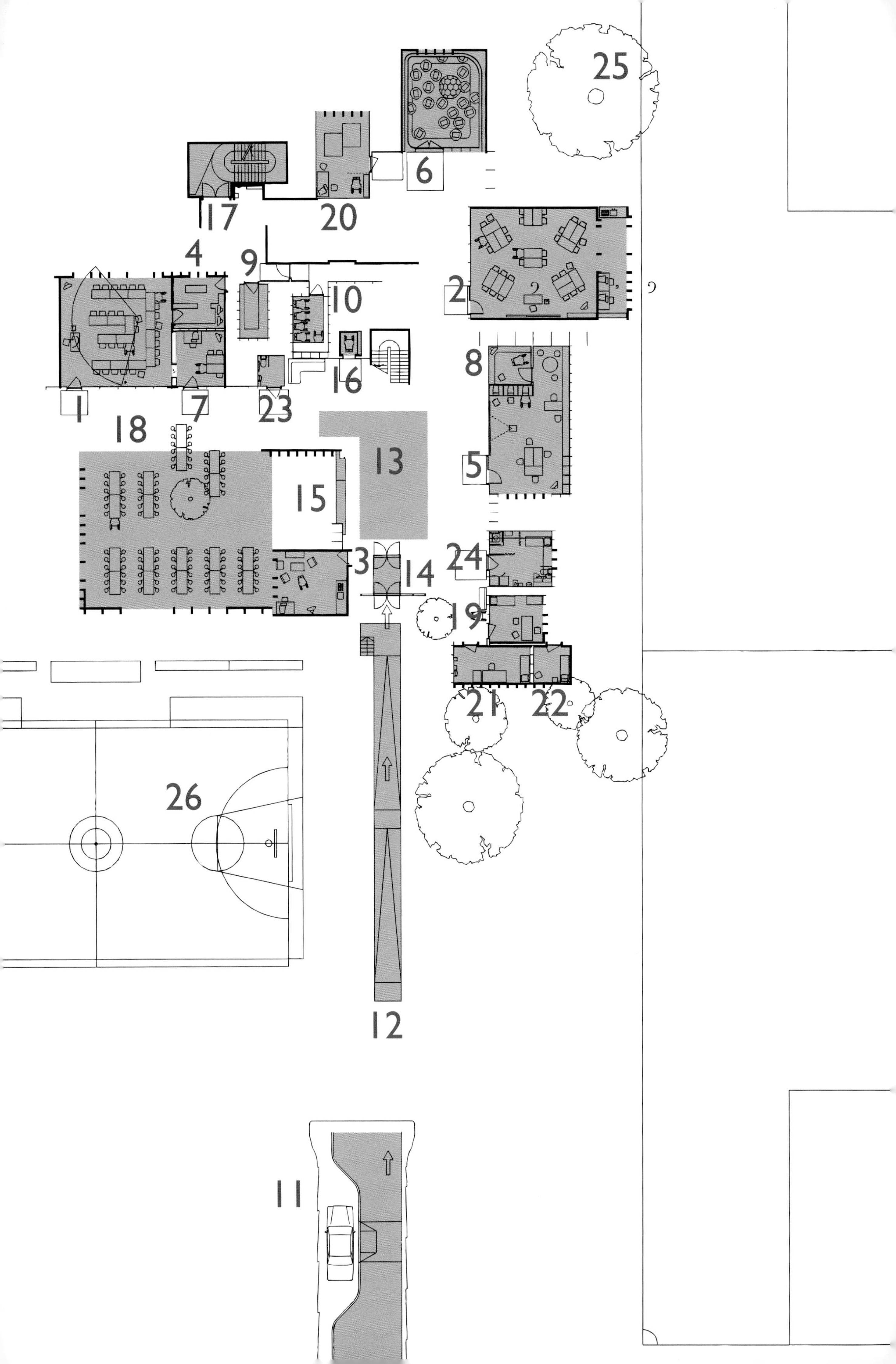
25
6
17
20
4
9
10
2
8
16
1
7
23
18
13
5
15
3
14
24
19
21
22
26
12
11

PART 3 Provision

Genuine inclusion needs to consider school life from the perspective of the pupil and seek ways to provide them with choice, dignity and self-esteem through the design of *inclusive learning environments*. The aim of this third part is to examine issues raised in Part 1 in terms of the implications for provision. It moves in scale from whole school issues to the examination of particular areas and rooms within school buildings and grounds.

3.1 WHOLE SCHOOL ISSUES

PLACEMENT

A school is essentially a place where pupils learn. They learn through gaining access to the curriculum and by having the chance to explore a rich array of personal experiences such as developing friendships or expressing themselves through music. The building supports the learning process and can aid, or hinder, the delivery of the curriculum and other less formal experiences.

Within the learning environment of the whole school are found a series of teaching areas, such as classrooms, small group study rooms, play areas and sports facilities. There are also non-specialist areas, such as corridors, hallways and outdoor courtyards. During the school day, pupils and staff move around the school in order to access a range of activities taking place in different areas.

Building Bulletin 77 'Designing for Pupils with Special Educational Needs: Special Schools', includes descriptions of additional areas such as sensory rooms.

The naming and placement of these different areas can be a fundamental consideration when becoming more inclusive, for a number of reasons. Take the example of *sensory rooms*. The name *sensory room* tells us a great deal about who is expected to use the space, the kinds of special needs they are assumed to have and, on a psychological level, their place as special within the school. Where it is positioned in relation to other activities will have practical implications. For instance, locating a *sensory room* in a separate building that pupils have to access via an external courtyard could make it physically difficult and time-consuming to get to.

Part 3.2 suggests area guidelines that allow for the accommodation of pupils with special educational needs and disabilities.

Becoming more inclusive involves thinking carefully about the flexible use of space and the creation of additional areas such as *resource bases*, storage areas for specialist equipment and all the other 'special' places that may need to be provided. This guide does not advocate one particular model. In some contexts, school staff, pupils and parents all value clear distinctions and physical distances between specialist teaching areas. In other contexts, it might be seen as appropriate to cluster *resource bases* with other teaching spaces. It is important to consider:

- the distances between activities and whether different areas will be easy to find;
- how pupils will move between, and gain access to, areas;
- the wider circulation flows of pupils between spaces;
- the spill-over of noise from different activities;
- how easy it will be to move specialist furniture and equipment between areas;
- the scope for changes to be made to the types of spaces provided and the scale of provision - the adaptability of the school environment. If, for instance, the capacity of pupils with a particular *sensory impairment* increases, will there be scope for increasing the number of *resource bases* within the school?

Many of these issues are discussed in relation to wayfinding.

In terms of more subtle issues to do with the messages we receive about ourselves from the way schools are organised, it is important to consider:

- the impression that pupils and others will pick up from the way spaces are described and placed within the school environment;

- the scope for spaces to be used by different groups of pupils pursuing different activities - what is described as the flexibility of areas within the school. Such multiple use during the school day or changing use over time may help break down imagined barriers between different pupils.

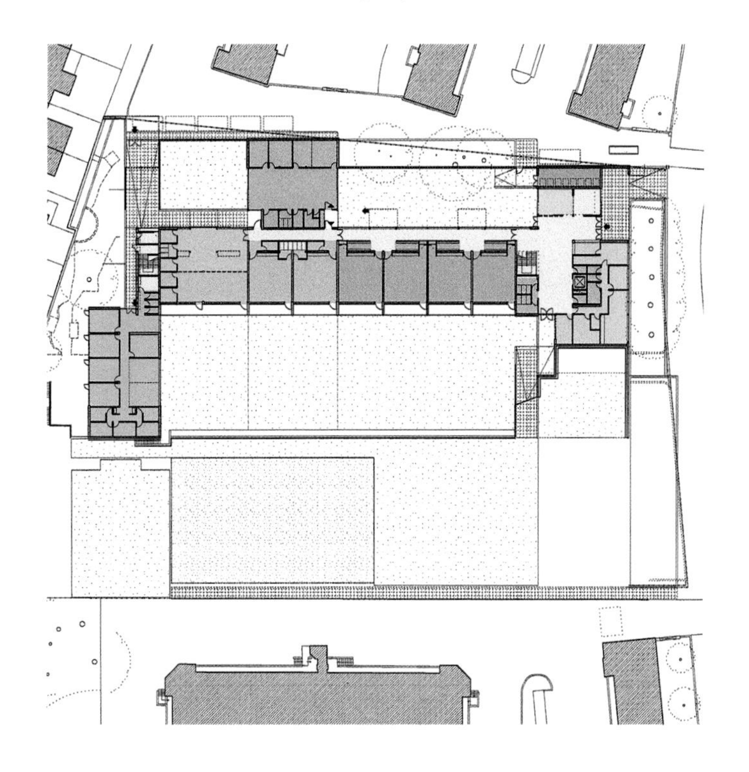

Plan of alternative locations for the resource provision, with each colour showing a possible location. © AHMM Architects.

PLACEMENT OF A NEW RESOURCE PROVISION

The process of incorporating a new resource provision into a *mainstream school* raises interesting design issues. New Tulse Hill Primary School is to be rebuilt, incorporating a resource provision for profoundly deaf children. The goal of including this provision within the main body of the school building, and including pupils within the day-to-day life of the school has taken its place as a core design criterion.

The design solution, however, is by no means straightforward. Should the resource provision be a building within a building? What are the acoustic problems that arise if pupils with *hearing impairments* are working within a noisy environment? How can specialist teachers access the resource provision easily? If it is located at the end of the main building, does this appear to be a form of segregation or simply a pragmatic and creative way to best meet with needs of all the school pupils? These are the questions the school has had to address.

WAYFINDING

Linked to questions of placement is the issue of how pupils can orientate themselves and find their way around school buildings and grounds.

Layout and landmarks

The layout of a school will govern how easy it is for a pupil to avoid getting lost. When deciding where to locate new facilities, it is important that:

- entrance areas should be visible and if there is an information desk it should be low enough for pupils in wheelchairs to see over;
- the distances pupils have to travel between activities and the complexity of the route are considered;
- noise levels, smells and other non-visual signals emanating from different spaces provide clues to help pupils find their way.

DESIGNING A SCHOOL LAYOUT

Great Binfields Primary School is a new build project in Hampshire, incorporating resourced provision for children with *visual impairment*. Being relatively small in size, the architects have developed a simple layout that will enable pupils with *visual impairments* to find their way. A main corridor runs the length of the school with teaching rooms on one side and support spaces on the other. This circulation space will be well-lit and care is being taken to design good internal acoustics and plenty of outdoor shading using trees and shelters throughout the grounds. All signs will be in appropriate text and incorporate Braille.

Architectural features can signal a change in activity at a subtle level, such as the use of colour or shelving as a partition that can indicate to pupils a point of arrival and the need for a change in behaviour.

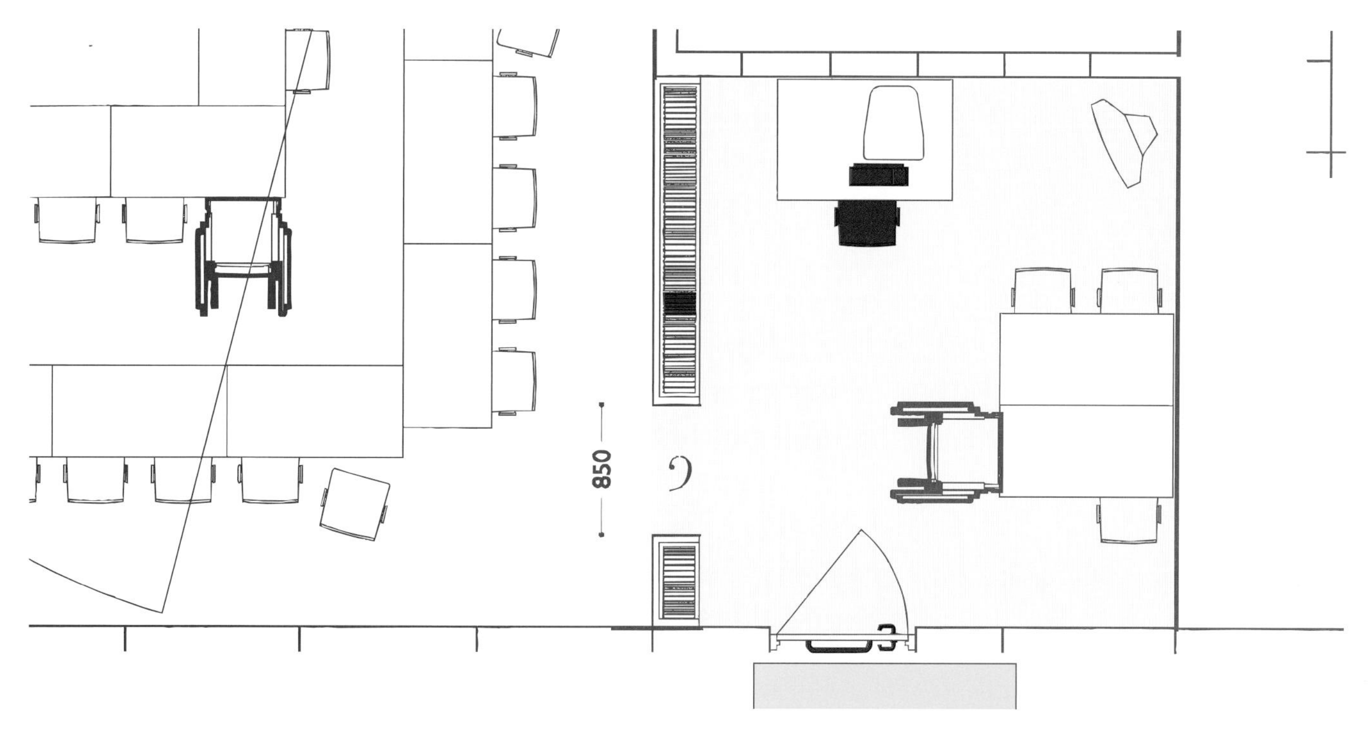

Social Spaces.
Small group area separated from the classroom by high bookshelves which act as an acoustic and visual divider. Access is available through the classroom and from the corridor.

Landmark features such as seating, display areas, and plants can all help pupils orientate themselves and move independently around the school. Views to outside and through to other notable parts of a building can also help pupils find their way.

The use of colour, texture, acoustics, lighting and signage can help pupils orientate themselves and journey through a school. Circulation through large open areas can be difficult for pupils, especially those with visual impairments. In such areas, it is important to define routes or areas with contrasting floor finishes and textures. At the same time, brightly contrasted and bold patterns to wall and floor coverings can cause confusion, especially for pupils with visual impairments, and so should be used selectively.

Signage

The art of designing signs and signage is rapidly evolving. It is worth considering the following issues.

The RNIB and the Joint Mobility Unit have collaborated in the publication of 'The Sign Design Guide'. Part 4.3 provides contact details.

- Make clear distinctions between signs that offer directions and those that indicate arrival.
- The design of signs should allow for contrasting colours, serif-free text, simple and consistent use of symbols and the provision of tactile information, including Braille.
- Signs need to be well lit.
- Signs need to be carefully located so as to help pupils throughout their journey by providing information at junctions or in long passageways.
- The use of voice activated signs may need to be considered.
- In addition, it is important to think about the height of the sign. There is no perfect height. Pupils vary in stature, and other pupils passing by may obscure what is visible to a pupil who uses a wheelchair. However, it is better to fix signs lower rather than higher.
- Some signs are not written. For instance, rails along corridors can be fitted with discrete buttons on the underside or bends in them to indicate that the pupil is approaching a door.

Examples of signs.

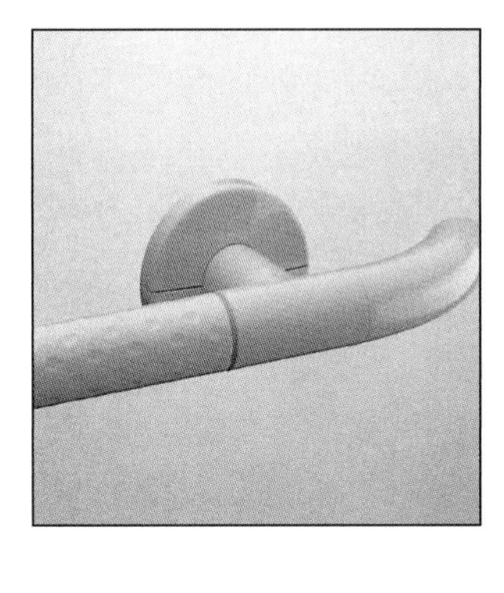

CIRCULATION

Building Bulletin 91 'Access for Disabled People to School Buildings' provides useful information.

Circulation within the school grounds

The third issue to address is how pupils gain access to the school from outside. Improving access to the school will also benefit younger siblings, parents using push chairs and other visitors.

- The organisation of routes to and around a school should not discriminate against pupils with *special educational needs* and *disabilities*. In principle, all pupils should be able to enjoy the same routes.

- There should be clear drop-off points close to the school entrance area with a well-marked, preferably covered walk-way supplied with hand rails leading to the main entrance area.
- Larger parking bays are required to allow pupils with mobility difficulties to get in and out of vehicles.
- Kerbs should be dropped between setting down points and the principal entrance areas.
- External routes and level changes should seek to minimise the effects of gradients.
- Main external routeways should be clearly signed, well lit and free from obstructions such as over-hanging branches.
- Ramped surfaces should be clearly indicated and all out-door surfaces should be non-slip. The gradient of ramps should be 1: 20 if no longer than 10m, 1:16 if no longer than 6m and 1:12 if no longer than 3m.
- There should be the option for pupils to use ramps or steps as some pupils will prefer to use steps.
- Paths should allow room for passing and have defined edges.
- Planting can provide valuable clues and orientation information.
- External steps should have consistent treads and risers with contrasting nosing, and handrails set at appropriate heights.

Detailed guidance for ramps and steps should be sought in Building Bulletin 91 'Access for Disabled People to School Buildings'.

A stair lift used for small changes in floor level, where stairs are wide enough and where interference with the main circulation route can be avoided.

The stair appears to be wide enough to accommodate a platform lift which would avoid obstruction of the hand rail for other users.

Circulation within the school buildings

When considering the layout of a new design or the adaptation of an existing school, it is important to think about how people will move around the school. What physical obstacles might bar movement? How can pupils with *special educational needs* and *disabilities* safely leave the premises in the case of an emergency? How can pupils orientate themselves and avoid getting lost? These are the key issues to address.

Horizontal movement

- Space should be allowed in all areas for pupils to pass each other. If narrower spaces cannot be avoided then doors into spaces off the main route should ideally have at least 1000mm doorsets.

ROOM TO PASS

Filsham Valley School found that 1800mm was not wide enough to allow pupils using wheelchairs to pass each other easily. The design therefore allows for corridors to be 1900mm or more at points around the route.

Building Bulletin 91 'Access for Disabled People to School Buildings' provides information on issues of physical access.

- Allowance for turning circles for wheelchair users should not be less than 1500mm in diameter.
- Any lobby should provide enough space for a pupil using a wheelchair to move clear of the first door and its swing before negotiating the second.
- Where double doors are required in any corridor, one leaf should always provide a clear minimum opening of 750mm.
- Routes should not be cluttered. Items such as coat hooks and fire extinguishers should be positioned or recessed to be clearly visible but not obstruct or create hazards for pupils.
- Floor coverings are important both in encouraging mobility and keeping people safe

in the event of a fall. For ease of wheelchair use, floor surfaces should be firm or hard, well-fixed, and non-directional. Some deep pile carpets can create problems for users of wheelchairs or mobility aids. Junctions between surfaces should not create a tripping hazard and should avoid visual confusion, for instance, by appearing to suggest a step where none exists.

- Floor surfaces should be slip-resistant, even when wet.

Vertical movement

Changes in level can be achieved in a number of ways including using stairs, ramps, lifts, platform lifts and stair lifts in certain situations.

- In all cases, it is important for the change in level to be clearly indicated so as to avoid pupils falling, stumbling or losing their balance. This can be achieved through the use of colour and tone, changes in texture, lighting and signs.
- Handrails should also be provided on both sides wherever possible when levels change.
- On stairs, nosings should be clearly marked in contrast with the treads and risers.
- Handrails should be provided alongside all stairs on both sides. If it is a flight of steps, rails need to be provided for a short distance on the levels at both the top and bottom to allow pupils to balance and rest.
- Different designs for the size and shape of handrails can offer protection and support or act as an indicator of changes to come. The design also needs to provide ease of grip for those with manual handling and dexterity difficulties.
- Great care needs to be taken with lifts. Chairlifts are not recommended for use in schools. Platform lifts can be used over changes in level within a storey, but platform lifts between storeys are potentially dangerous and would require some form of enclosure.
- Stairlifts are generally not recommended but may be

the only solution in an existing building if there is no space adjacent to the stair, or there is not enough width for conversion to a platform lift leaving adequate space to one side.

Doorways and doors

Ill-designed doorways and doors can often be barriers to movement.

- Clear door openings should be at least 800mm wide.
- Thresholds should be flush wherever possible.
- In consultation with fire officers, ways should be found to avoid doors wherever possible. Fire doors should be held open by alarm-linked devices wherever possible.
- Door frames and doors should be clearly contrasted to help pupils, especially those with *visual impairments*, distinguish them from their surroundings.
- Where needed, doors should be carefully weighted so as to be easy to open. Push button mechanisms may be needed where doors are heavy and time should be allowed for a slow-moving pupil to pass through.
- Door furniture should be easily gripped and operated, clearly visible and contrasted against its background.
- Visibility panels should be provided at different levels to allow young pupils, pupils using wheelchairs and standing pupils to see into a space. It is important to note that for many children visibility panels can be dangerous without the use of safety glass.

Building Bulletin 91 'Access for Disabled People to School Buildings' provides essential information.

Safe escape

Concern about child safety, vandalism and the protection of equipment have all brought the issue of security to the fore. Some measures, such as securing all exits, including fire exits, can give rise to conflicts between safety, security and physical accessibility. Some of these conflicts can be overcome by introducing fixtures such as door bells and lever operated doors. Modified management procedures can also help. What is clear is that the issue of safe egress in the case of fire needs careful planning and detailed consultation with fire officers.

Procedures

- Any security arrangement for the school should be checked for accessibility. For example, arrangements for the use of swipe cards or coded entrances to parts of the building should be chosen and located to maximise access.
- It is important to incorporate the needs of all pupils within the school into evacuation plans and procedures. Can pupils see, hear and understand fire signals?
- In some cases, pupils may temporarily have to wait in refuge areas before being helped to a place of safety. Refuge areas must be provided at all escape stairways, one on each upper or lower level, and the width of stairways should allow for wheelchair evacuation. Those waiting in refuge areas must also be able to communicate with those organising the evacuation using some form of intercom system.
- Some pupils will need assistance from an adult to evacuate the premises safely. The nature of this assistance, the roles to be played by different members of staff and necessary training should be established, confirmed in writing and tested regularly.
- Evacuation chairs can be suitable for some pupils but not all.
- Management issues, such as keeping corridors unobstructed, testing alarms and maintaining door closers are all crucially important to ensuring safe egress in the case of fire.

FIRE PROCEDURE

Walney Secondary School in Cumbria has developed procedures for safe egress which include the provision of evacuation chairs at the top of every flight of stairs, which were provided by the Local

Education Authority. A member of staff is allocated to every child with mobility problems and staff are also trained to assist those with sensory disabilities.

Fire alarm system

Building Bulletin 91 'Access for Disabled People to School Buildings' and Building Bulletin 77 'Designing for Pupils with Special Educational Needs: Special Schools' both provide guidance on safe escape.

Managing School Facilities Guide 6 'Fire Safety', DFEE 2000.

- Those working with schools should develop fire alarm systems in consultation with the fire authority.
- BS 5834 fire detection and alarm systems for buildings gives the recomended distance between manual break glass 'call points'.
- Signage of call points needs to be clear.
- Alternative alarm signals may need to be considered, such as visual alarms, paging systems, vibrating signals or sound signals within carefully selected frequency bands.

ENVIRONMENTAL DESIGN

Lighting and colours

It is difficult to overstate the importance of good lighting and colour contrasting within school buildings and grounds.

- Good levels of natural light should be created wherever possible.
- Lights should be positioned where they do not cause glare, reflection, confusing shadows or pools of light and dark that can be misleading, especially for pupils with *visual impairments*.
- If possible, all lighting, whether natural or artificial, should be controllable and adjustable to suit the needs of individual pupils. It may not be possible to install lighting controls in all spaces. Although some lighting systems allow them to be added later when needed. This allows the lighting to be adjusted and controlled to suit the needs of individual pupils.
- Attention should be paid to the lighting of potentially hazardous areas such as stairwells.
- Sudden changes in light level should be avoided as it can be disorientating, especially for pupils with *visual impairments*.

Building Bulletin 90 'Guidelines for Lighting Design for Schools', Building Bulletin 91 'Access for Disabled People to School Buildings' and the RNIB publication 'Building Sight' all provide advice on lighting, colours and colour contrast.

- Wherever possible, glare should be avoided. Glossy wall and floor surfaces give rise to reflections, which can be uncomfortable and cause visual confusion.
- Lighting can be used to enhance subtly the impact of variation in colour and contrast, thus providing visual clues to help pupils orientate themselves and find their way around.
- Up-lighting, set above standing eye level, can be especially helpful in creating a glare-free environment.
- Mains frequency fluorescent lighting can create a magnetic field, which can cause a hum in hearing aids. High frequency fluorescent lighting should be selected to eliminate this problem but you need to check with hearing aid manufacturers that this is the case.
- Careful selection of contrasting colours can help differentiate parts of the school. Examples include the designation of different floors and specific teaching areas using different colours.
- When considering colour schemes, there is no need to adopt strong contrasts that create garish interiors. Most pupils with *visual impairments* can perceive relatively subtle contrasts. Using relatively small areas of strong colour and large areas of light colour tends to make the most efficient use of natural light and lighting.
- The use of contrasting colours also helps pupils identify areas, doorways, electrical switches and other furnishings. The key area of interest to pupils with *visual impairments* is the space seen when looking downward and within two metres of where they are standing.
- Pupils with *visual impairments* often use the ceiling area for orientation and to identify the size of the space they are in. Adequate contrasting is therefore needed between ceilings and walls.

Further information can be found in RNIB book 'Building Sight'.

Acoustics

The aim is to control noise levels, frequencies and reverberation times. The needs of pupils with *hearing impairments* should be considered. Acoustic conditions also have a profound influence on the

Building Bulletin 87 'Guidelines for Environmental Design in Schools' provides detailed information on enhancing the acoustic quality of school environments.

lives of pupils with *visual impairments*, who use sounds to help orientate themselves, as well as pupils who find loud noises distressing. Indeed, all pupils need good acoustic conditions to help them concentrate and learn. The following core points should be considered.

- When assessing where to locate a building, extension or activity it is important to note the proximity of the site to external noise sources (such as roads and playgrounds) as well to other noisy activities such as food preparation or music.
- Noise levels can be controlled by the use of buffer zones and physical barriers, such as walls and windows.
- Materials need to be carefully selected to provide sufficient acoustic absorption in a space. The amount of absorption affects the reverberation time. Provision of curtains, acoustic ceilings, carpets and pinboards all increase the absorption in a space and therefore reduce the reverberation time.

MANAGING NOISE LEVELS

Westgate Secondary School is a 'bi-lingual' school that accommodates pupils with *hearing impairments* who use both speech and sign language. When converting an existing classroom into a resource base for pupils with *hearing impairments*, several issues had to be addressed. These included reducing noise levels from rain drumming on the roof as well as improving the sound insulation of the walls. Good lighting was also important to help pupils to see each other signing and lip reading and to see teaching and support staff. A small dedicated room with a high level of sound proofing has also been established for specialist support by therapy and advisory services.

- The shape and proportions of a space will affect the acoustic quality. Large halls, if incorrectly designed,

The Royal National Institute for the Deaf (RNID) produces a wide range of publications on the experiences and design needs of people with hearing impairments.

may produce pronounced echoes making it hard for all pupils especially those with *hearing impairments* to understand speech. High ceilings will create very different acoustic qualities from a low-ceiling space. L-shaped rooms may contain dead spots where it is hard to hear sounds intelligibly. These problems can be addressed by changing the shape of the room, by using sound absorbent materials and through the careful positioning of diffusive materials.

- Open plan areas can cause difficulties in terms of controlling acoustics and should be avoided wherever possible. In some cases, intrusive noises can be masked by other calming sounds such as music or the sound of water.

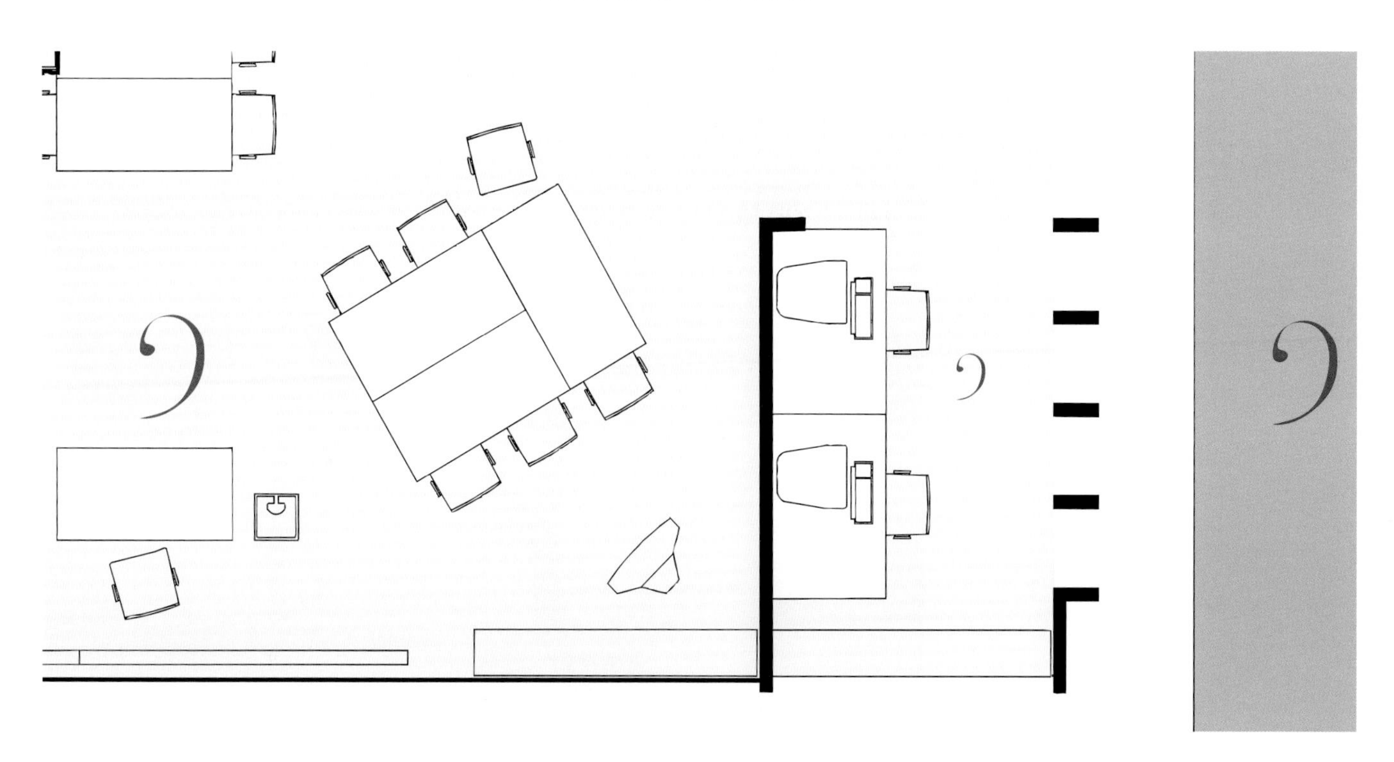

A building's location and fabric should aid acoustic performance and reduce external interference, allowing the teacher's voice to carry.

Heating and ventilation

- Heating and air conditioning units should be as quiet as possible and regularly maintained in order to keep noise levels low. If possible, air conditioning should be avoided.
- Heating and ventilation systems should be controllable and adjustable, according to the needs of individual pupils.

Building Bulletin 87 'Guidelines for Environmental Design in Schools' provides information on heating and ventilation systems.

Storage

Storage is an extremely important whole school issue.

Space is needed for storage of additional equipment and teaching resources throughout the school. Extra space is needed:

- in entrance areas for storage of *mobility aids*, wheelchairs and other equipment;
- for the storage of pupils' personal possessions, such as food, communication aids or clothing. These areas and the storage furniture need to be carefully designed to ensure that all pupils can use them safely and independently;
- within teaching rooms, sports areas and small group rooms;
- in office areas for the storage of information and records;
- for technicians to store specialist equipment such as *mobility aids* and batteries.

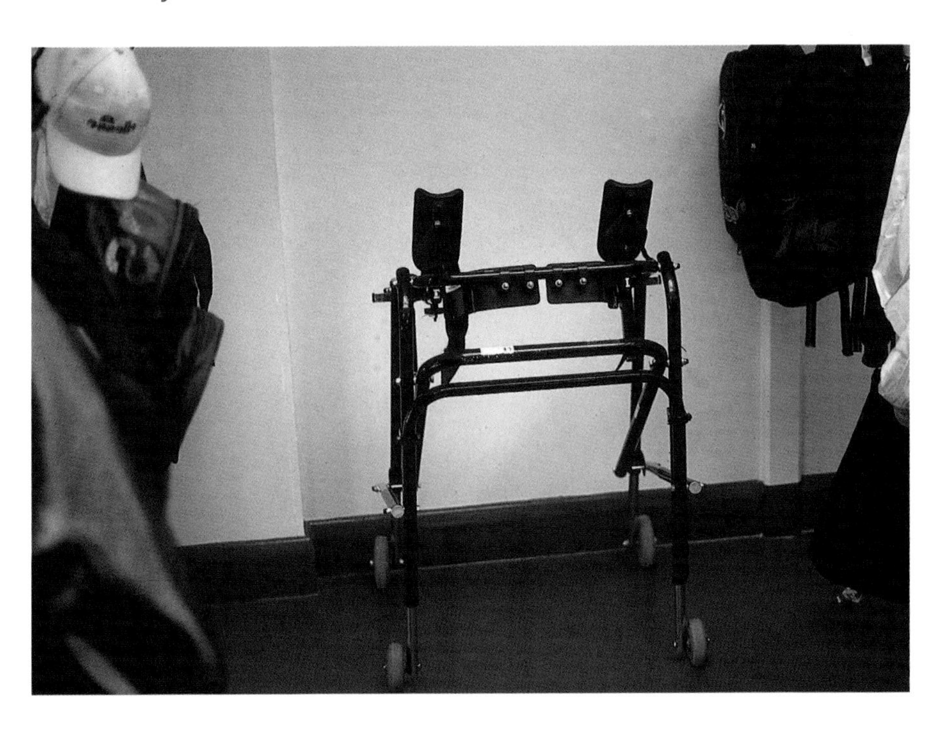

A roller walker.

THE NEED FOR STORAGE SPACE

In planning for storage space to accommodate equipment used by pupils with physical disabilities, staff at Pennyman School in Middlesbrough drew up the following list of bulky items: wheelchairs; wheelchair batteries; roller walkers; standing frames; and adjustable chairs.

FIXTURES, SERVICES AND BUILDING ELEMENTS

Electrical fittings and controls

Provision may need to be made for battery recharging in a well-ventilated area.

- The main electrical supply to a school generates a considerable magnetic field that can cause a loud hum to all those using hearing aids. If possible, main cables should be routed away from areas used by pupils and other people.
- Computers, overhead projectors and lighting can create background noise and can interfere with hearing aids.
- *Induction loops, radio systems, infrared systems* and *sound field amplification systems* may need to be fitted in areas where pupils with hearing aids are likely to receive information. Advice should be sought from manufacturers on selecting and installing loop systems.
- Sound can 'spill' out of the field of an *induction loop* and there can be overlap where loop systems are located near each other. This may make a loop system inappropriate where confidentiality is required: an *infrared system* may be more suitable. *Radio systems* may now be considered for use in the classroom.
- There should be an adequate supply of electrical sockets throughout the school. Their height, colour and precise location should be carefully planned.
- The installation of occupancy controlled lighting, automatic flush toilets and other electrical equipment may need to be considered.

USE OF INFORMATION TECHNOLOGY

Vickerstown Primary School in Cumbria avoided problems of leads from floor sockets and bulky furniture by providing pupils with battery powered lap-top computers. The pupil using the computer is thus able to work in different positions within the teaching space.

Building Bulletin 91 'Access for Disabled People to School Buildings' and Building Bulletin 87 'Guidelines for Environmental Design for Schools' provide useful information.

- Amplification equipment may need to be installed.
- Large electrical switches can provide for the possible manual dexterity needs of current or future pupils.
- Alarms and communication systems may need to be installed for the pupils and adults working in the school.

Ergonomic design of furniture

Providing comfortable furniture that allows the pupils to participate in educational and social activities is of the utmost importance. However, there is no general measure from which the design needs of pupils at different ages can 'read off'. Instead, the challenge is to create furnishings that are easily adaptable and thus responsive to the needs of individual pupils.

An adjustable worksurface

Building Bulletin 91 'Access for Disabled People to School Buildings' provides information on electrical fittings and controls.

In some cases, specialist furniture will need to be bought from manufacturers. In others, high-street furniture such as quality tested office furniture or beanbags can meet pupils' needs. When thinking about furniture, it is important to bring in health colleagues and other specialists to pool ideas. Other issues to consider include:

- the quality and robustness of furniture;
- the height of displays, mirrors, signs, light switches, shelving and other wall fixtures;
- the height of counters, desks, computer workstations, benches, sinks and other furniture;

- the area and design of work surfaces. Is there room for the pupil to arrange any communication aids they need? Is there room for an assistant to work alongside the pupil? Is the pupil able to reach keyboards and other learning tools?
- the style of the furniture and the kinds of messages it gives about the status of the pupils using it;
- the location of seating along routes, in outdoor areas and within teaching spaces;
- the detailed design of seating, and the provision of a mixture of seating with and without arm rests;
- the amount of space around furniture to enable pupils in wheelchairs or with mobility equipment to be comfortable;
- ensuring that colour and texture are carefully selected to enable pupils to locate and identify furniture and equipment;
- the finishings of furniture and the need for exposed edges to be rounded;
- the width of counters and work surfaces and distance to objects or fittings that the pupil needs to reach;
- grab handles in appropriate places, such as in the toilet and to enable a pupil who has trouble standing, at a sink or work surface;
- easy access to hand washing facilities for pupils with poor hand co-ordination.

The use of a contrasting coloured mat helps to identify the position of a hot teapot.

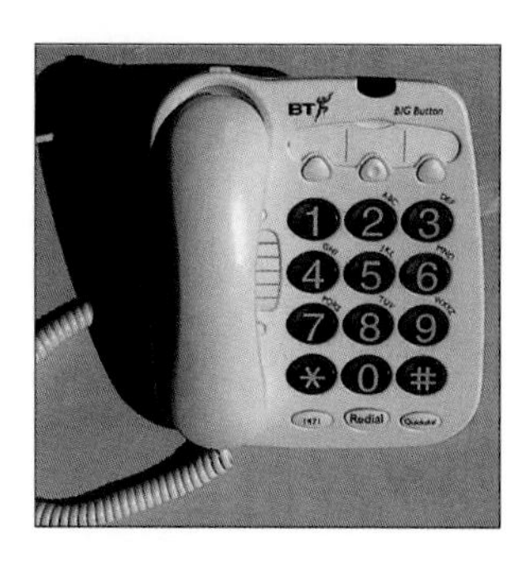

An example of specific design, a telephone with large keys.

3.2 SCHOOL SPACES

The aim is now to move through a school and examine the particular design issues that relate to different areas and rooms within it.

TEACHING ROOMS

The number, type and clustering of teaching spaces will vary from school to school. In some cases, all of the school will be shared by all of the pupils. As explored in 3.1, in other cases most of the rooms dedicated for use by pupils with *special educational needs* and *disabilities* will be linked to, but separate from, other teaching spaces. In all cases:

- it is important to provide a variety of different sized teaching spaces;
- it is useful to link large teaching spaces to smaller teaching spaces where assessment, support teaching, different kinds of therapy and small group study can take place, and specialist equipment can be stored;
- there should be extra storage space for group and individual teaching aids and equipment.

CREATING QUIET ROOMS

Cornwall County Council promotes the idea of creating support rooms within mainstream schools where a child can work in different surroundings and away from circumstances that might have been causing a problem. This facility is particularly useful for pupils with *autistic spectrum disorder* and *emotional and behavioural difficulties* who are then given a chance to review with a teacher the nature of the problem they face and how to deal with it in the future.

- Teaching spaces, wherever possible should be adaptable and flexible. Adaptable spaces are those that can be easily changed structurally. Flexible spaces are those which can be easily rearranged and thus be used for different purposes and by different groups of pupils.

- Consideration of furniture design and layout can help create flexible teaching spaces and allow a mixture of academic and practical work suitable for pupils with different learning styles.

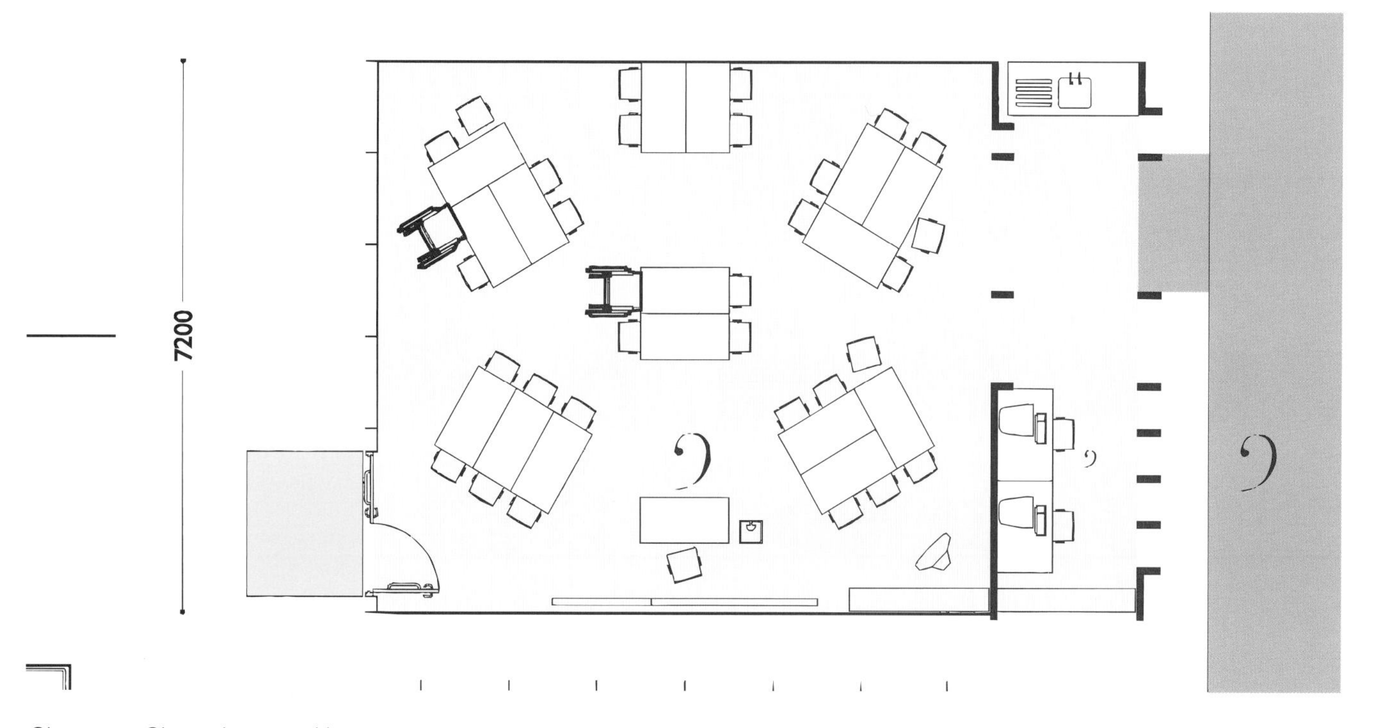

Classroom Cluster Layout with a practical/IT space attached to it. Alternative classroom arrangement allowing a pupil in a wheelchair to sit anywhere in the room.

- Teaching spaces should have some space for practical activities, quiet study or where a pupil can withdraw if he/she feels uncomfortable or need to work alone at his/her own pace.
- There should be additional room within the teaching space for teaching assistants to work alongside pupils or for wheelchairs to be parked.
- All pupils should be given a choice of places where they can sit and study: creating a 'special' place can be stigmatising and limit the range of activities a pupil can enjoy.
- There should be uncluttered routes through teaching spaces, which are safe for pupils with *visual impairments* to follow.
- There should be good lighting on the teacher's face to help pupils with *visual* and *hearing impairment*.

LIBRARY FACILITIES

Within schools, many rooms are designed for specific teaching activities, such as science, food technology or information technology. In each case, particular issues will arise relating to the use of equipment, design of furniture and so forth. Libraries also present particular challenges. Issues to consider include the width between shelves, shelving height, the height of information desks, lighting and signage. The way pupils access resources and actually handle books and equipment will also need to be addressed.

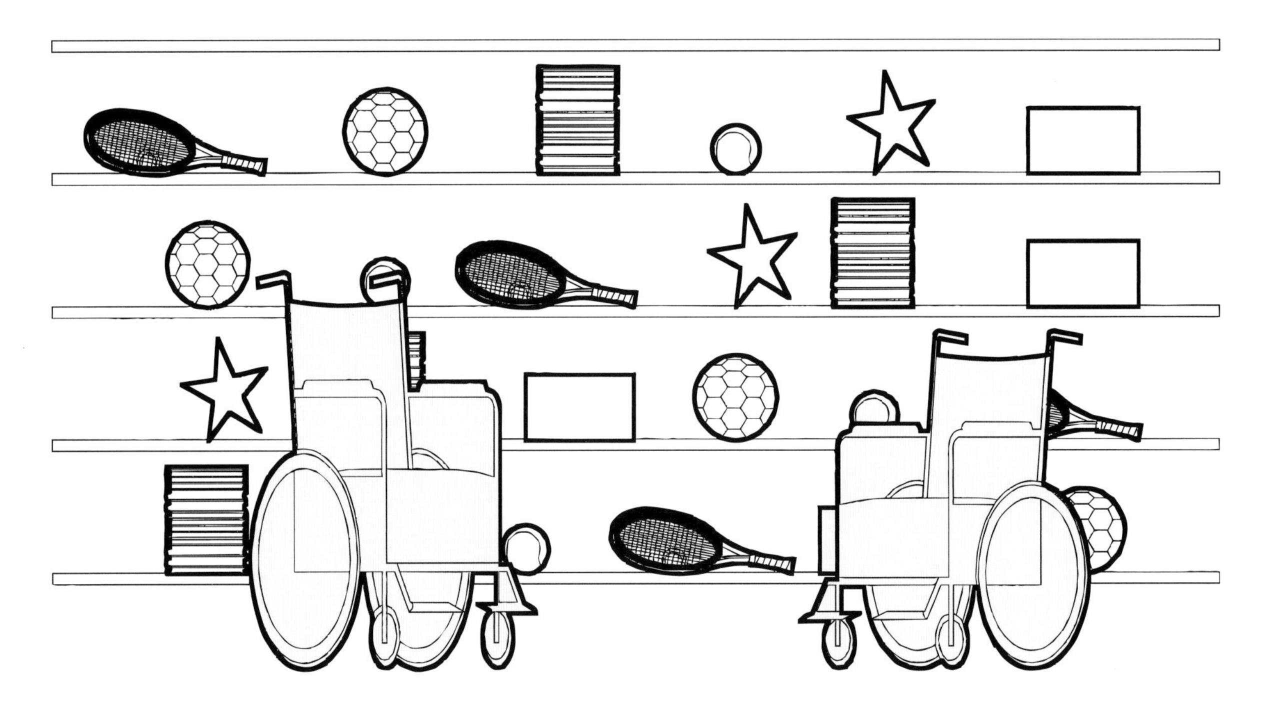

Arranging the disposition of things vertically on shelves ensures accessibility from a range of heights.

FLEXIBLE NURSERY CLASS BASE

Alexandra Special School was formed when an existing special school and a mainstream nursery were combined. It accommodates pupils with a very wide range of needs including physical *disabilities,* moderate *learning difficulties,* and *emotional and behavioural difficulties.* The nursery class base has also been designed for flexibility. It includes a large octagonal central space connected to bays for

The DfEE publication 'Furniture and Equipment in Schools: a purchasing guide' provides useful general advice on furniture design.

different activities, including: covered outdoor space; entrance to the reception area; toilets and changing area; kitchen; physiotherapy and speech therapy area; indoor and outdoor storage; and computer area. Two of the bays have movable walls so they can combine with the central area and the kitchen area has a stable door, which enables staff and pupils to see each other.

The recent DfEE publication, 'Designing for 3 to 4 Year Olds', provides advice on the design of nursery spaces.

SUPPORT ROOMS

Small group rooms that are designed to allow for flexible use are an invaluable resource for most schools. In some contexts, it may be possible to use such a room for a range of different functions. In others, a series of rooms may be needed. They could be used for:

- one-to-one or small group teaching work, counselling and therapy sessions;
- teaching and support staff who need to prepare lessons and carry out assessments;
- visiting therapists to store equipment, records and personal possessions and to work with pupils;
- parents and carers to meet with school staff and visiting therapists.

PARENTS' ROOM

Wigton Infant School accommodates children with a wide range of *special educational needs* and *disabilities* and the school staff work closely with local authority personnel as well as with parents. The decision was made to create a parents' room that could serve as a place to have events for parents, store reading material and other useful information, and hold discussions between parents, teaching staff, visiting professionals and pupils.

Additional space requirements for the inclusion of pupils with *special education needs* and *disabilities* include *SEN* support spaces predominantly for pupils with *special education needs* to access, but also of some use for all pupils in the school, as well as specialist classbases or support centres. These may be used for the majority of the school day by those with particular difficulties, such as a class for pupils with *hearing impairment* or facilities for those on the roll of a *co-located* or nearby special school.

The truly inclusive school will allow social, functional and locational inclusion of all pupils with *special education needs* on roll. Specialist classbases are likely to be only a temporary measure as the school progresses towards total inclusion. Such spaces, which may include specialist practical or vocational skills areas adjacent to classbases, would be in line with guidance given in BB77.

Technician's room 10 – 20m^2

This may be needed for adjustments, maintenance and repair of aids if the number of pupils using aids is high.

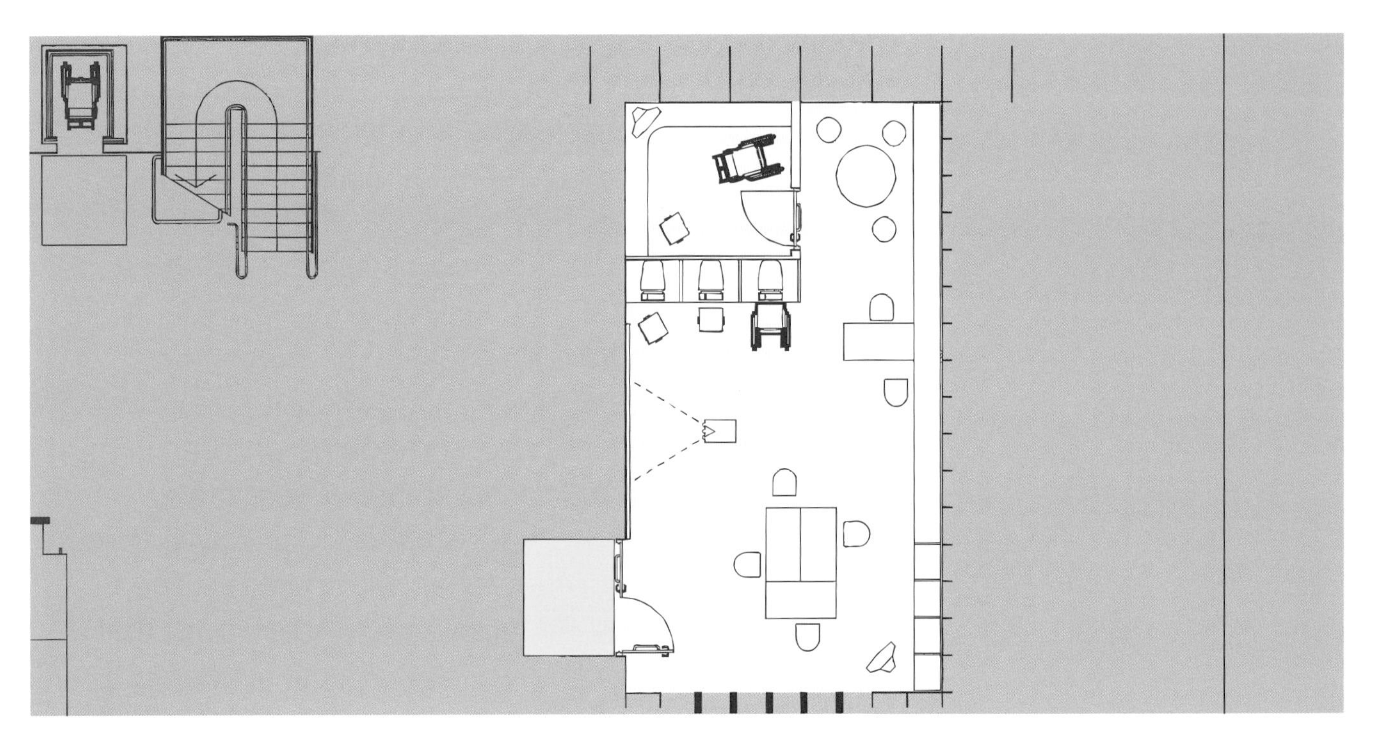

An SEN resource base.

SEN Resource Base 25 – 54m^2

These are key spaces for individual or small group teaching out of the standard classroom, for specialist materials and equipment, and for case conferences with staff and advisers. This room will often have an *SEN* coordinator (SENCO) and storage space adjacent, as part of a larger suite.

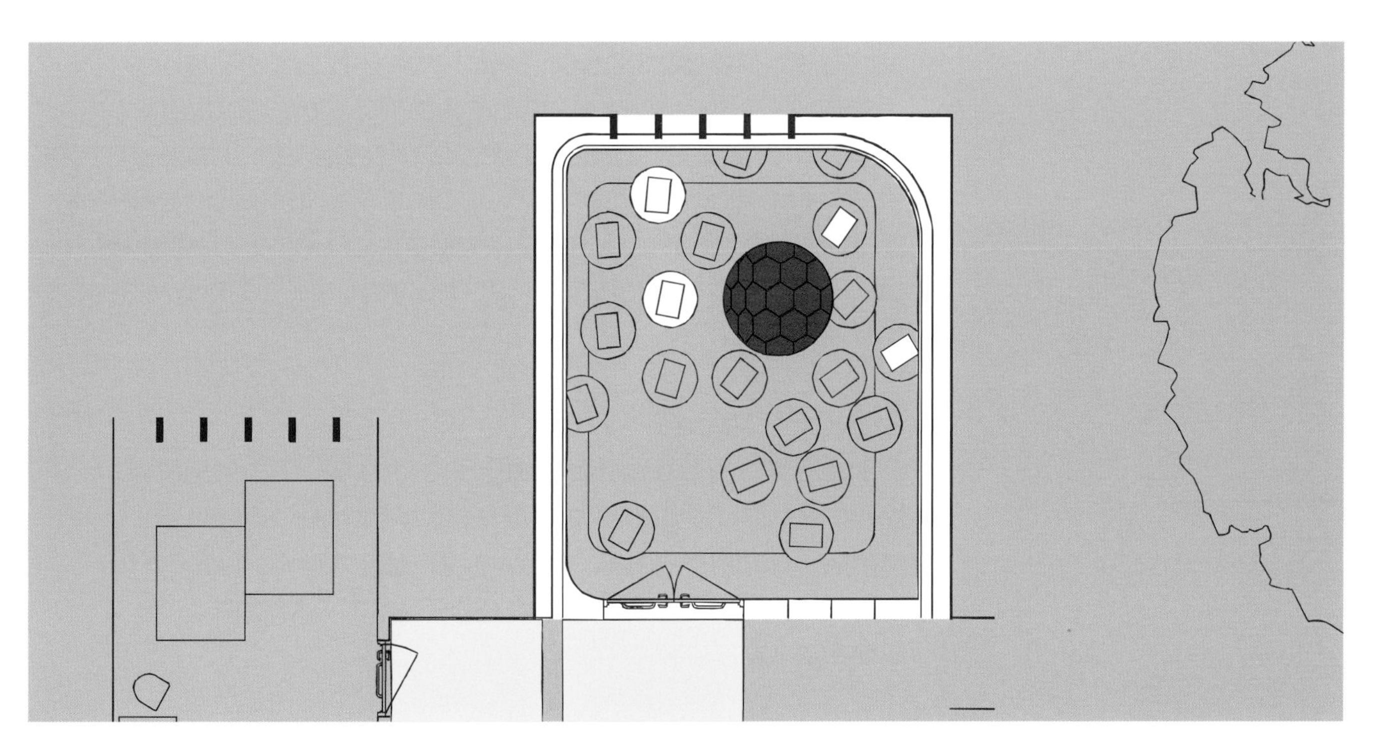

A soft-play area.

Soft-Play Area 10 – 30m^2

This should have low horizontal and vertical surfaces covered with foam padding and various other loose soft shapes, to allow the physical conditions for safe play, including that of a more lively and robust nature. Although it is particularly relevant for younger pupils, (and those with motor difficulties), it can be well used as a teaching medium (shape, dimensions, colour) for all pupils. It can also be useful for recuperative and assessment purposes.

Small Group Room 25 – 30m^2

If there is a sufficient demand, social spaces can provide an area where those who may be vulnerable or in need of additional support can develop social skills in small groups (usually of up to six) in a less formal atmosphere than the classroom.

Warm Water Pool 70 – 150m^2

Pools are increasingly being included in special schools which may be *co-located* but are unlikely to be cost effective unless there are sufficient numbers of pupils on roll that will benefit from hydrotherapy, or a high demand from the community.

See BB77 'Designing for Pupils with Special Educational Needs: Special Schools', 4.11- 4.12.

Sensory Room 12 – 20m²

This can be fitted with specialist equipment and black-out, to provide a range of experiences, including sight, sound, smell and touch, which are of value to pupils with severe *learning difficulties, sensory impairment* and some physical *disabilities*. The furniture and decoration of sensory rooms needs to be carefully planned. In many cases, stimulating design features such as soft furnishings can be introduced elsewhere in the school.

SENCO Office 6 – 10m²

The *SEN* co-ordinator may use the *SEN* resource base as an office or may need a further separate space, usually adjacent, which can also be used for private interviews. The area allows for secure storage of personal records.

SEN Central Store 5 – 8m²

For communication aids, records and equipment. It may be adjacent to the *SEN* resource base and/or SENCO office. There may be more than one if there is a high demand.

Building Bulletin 77 'Designing for Pupils with Special Educational Needs: Special Schools' offers advice on support rooms.

Wheelchair and Appliance Storage 8 – 10m²

Battery operated equipment may need to be recharged, so appropriate power sockets may be necessary. Pupils or staff with disabilities may need to use more than one wheelchair, and these will be stored here.

Building Bulletin 91 'Access for Disabled People to School Buildings' provides information on entrance areas.

ENTRANCE

- Entrance areas should be easily distinguished by their design, location and lighting.
- Signs, including tactile signs, should be used to mark the entrance area. Landmarks and design features such as planting, seating and tactile paving should be arranged to provide obvious route ways.
- Routes to the entrance area should be accessible and outdoor surfaces suitable for wheelchair users.
- Door furniture should be easy to grip and operate, and the force required to overcome the power of the door-closer should be kept to a minimum.

- Thresholds should be flush and, if possible, absolutely level.
- In terms of floor surfaces at exterior doors, a firm and flush entrance mat should be provided, and care should be taken to avoid tripping hazards.
- Power-operated automatic doors may be appropriate in some circumstances where doors have heavy traffic and both accessibility and energy conservation are considerations. Automatic doors that swing towards the user can be hazardous and should be appropriately signed. Automatic doors should not be able to close on a person.
- If necessary, the entrance should offer a transition zone where pupils with *visual impairments* can adjust from a bright exterior to a more subdued interior space.

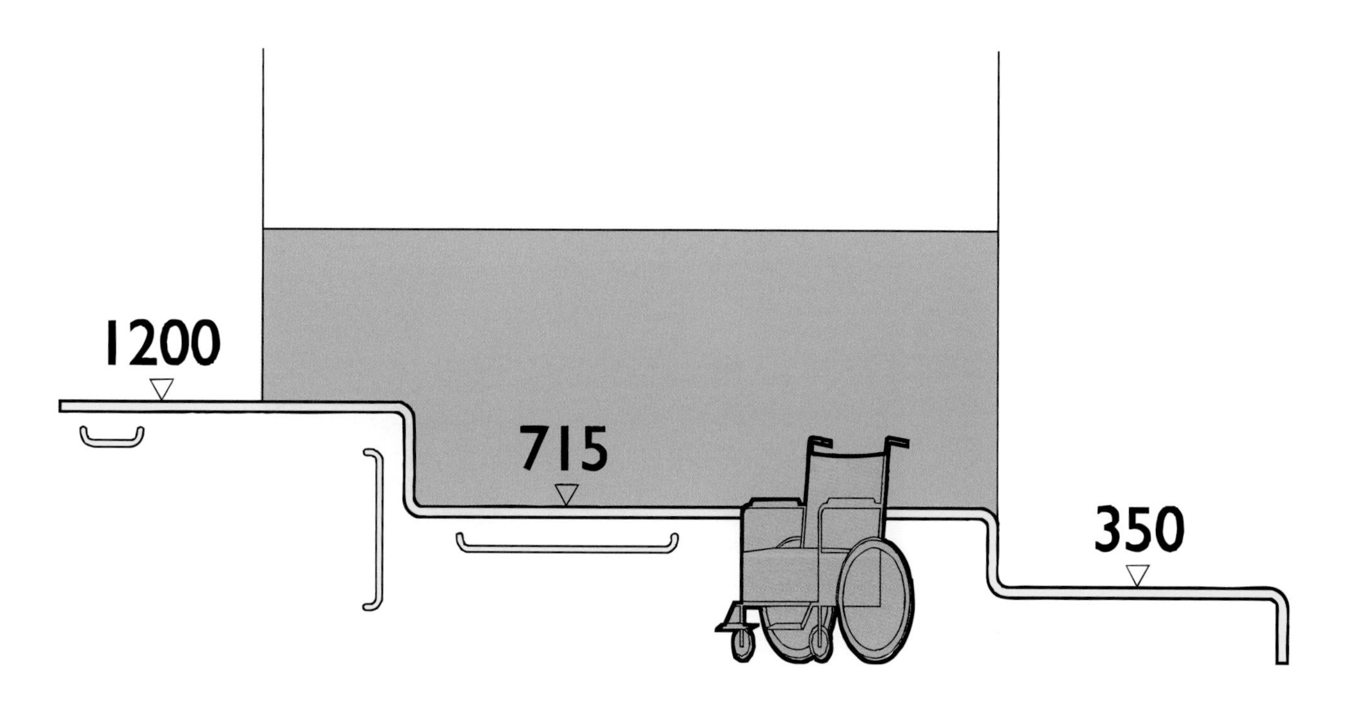

A reception desk with different level surfaces, a place to sit and a place to lean.

- Any reception point or information hatch should be clearly signed, have a lowered section, be well-lit and provide hearing assistance. Downlights should be carefully covered over receptions desks where pupils with *hearing impairments* may need to lip read.

- Waiting areas should be generous with room for parents, carers, visiting therapists, and space for wheelchair users.
- Storage room should be provided for wheelchairs. There should be adequate room for pupils to store their personal possessions and equipment safely.

DINING

There is a growing interest in dining areas being places where pupils learn more about nutrition, develop good eating habits and have the chance to relax. Yet, many tend to be noisy places where many pupils face difficulties in eating or socialising in a relaxed way. The situation can be especially hard for pupils who have special nutritional requirements or who have difficulty eating by themselves. Great care is needed in designing, decorating, furnishing and equipping dining rooms.

- All pupils should be able to eat in a dignified way with their peers if they so choose.
- Extra circulation space may be required to allow for pupils using wheelchairs, or who have mobility impairments.
- Additional space may be required to provide for parking space for wheelchairs.
- Space and equipment may be needed for the preparation of special foods and storage of equipment.

- Space may be required for assistants to sit alongside pupils and help them eat.
- The design of furniture, such as seats, tables and serving hatches should take into account the needs of all pupils. As one example, the height of serving counters should take into account the needs of pupils using wheelchairs.
- The choice of dining equipment, such as cutlery and plates, should take into account the needs of all pupils. For instance, avoiding metal cutlery containers can help reduce noise levels.

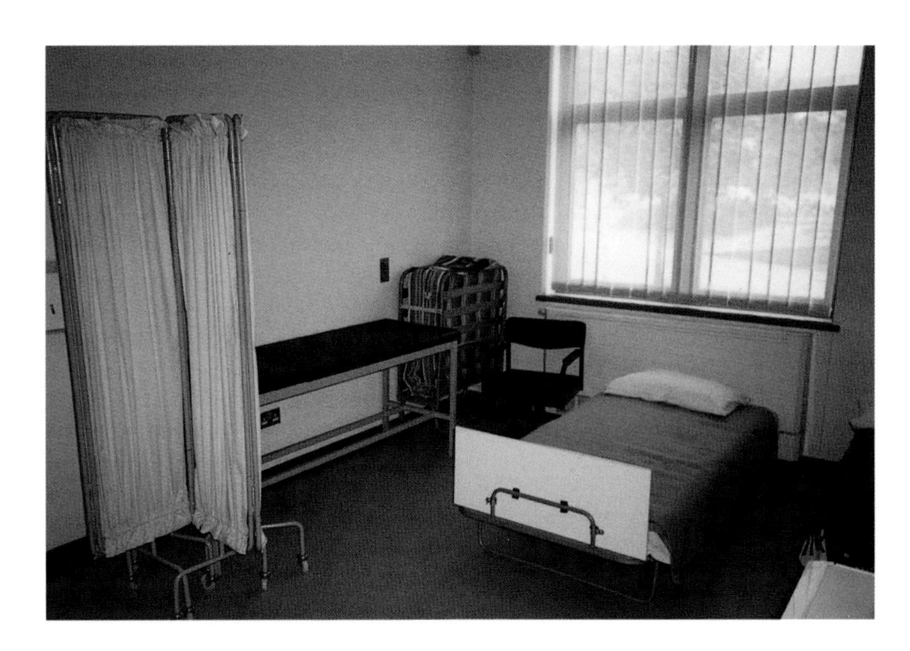

A school medical room.

MEDICAL AND THERAPY

There is a range of additional medical and treatment rooms that may need to be developed. Medical rooms should have adequate space for:

- pupils to move around freely, park their wheelchairs, be examined in comfort and privacy and take medication or other treatments;
- assistants who may be needed to help the pupil manoeuvre themselves;
- medication, the storage of records and information and additional equipment such as hoists and showers.

There may also be a need to design or adapt rooms so that pupils can have different forms of treatment and therapy such as physiotherapy, hydrotherapy and music therapy. Discussion should take place with relevant professionals and manufacturers on the specific design requirements and furnishings needed.

Other visiting specialists such as a speech and language therapist, occupational therapist or educational psychologist will generally be able to use the medical room or a small group room although records and equipment should be made readily available. If the number of peripatetic specialists is high, one or more further rooms may be needed.

The DfEE publish guidance 'Supporting Children with Medical Needs'.

Medical Inspection Room 10 –15m²

All schools must have an MI room, under the Education (School Premises) Regulations 1999, as a rest room for a sick pupil and for visiting specialists. In larger schools, or where there are a high number of pupils requiring medical care, there can be a medical suite including a nurse's room, or treatment room, a separate rest area and toilet/hygiene facilities.

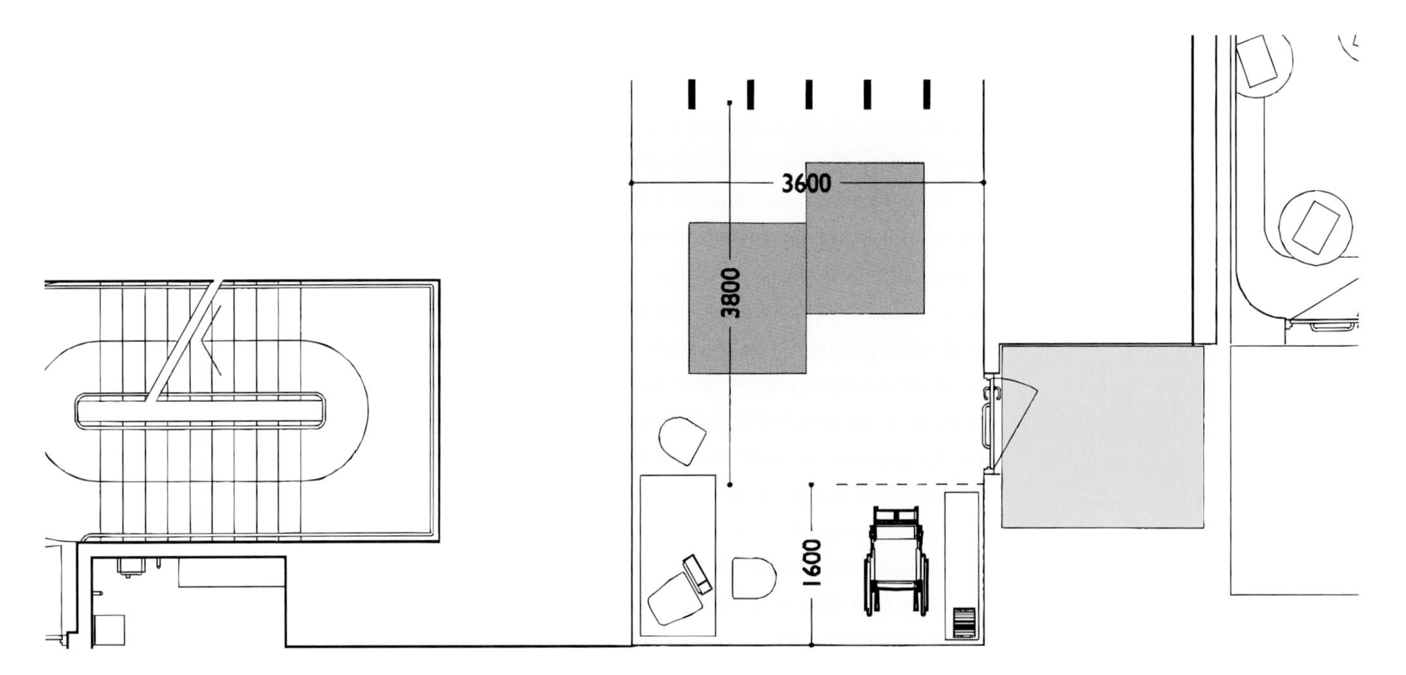

Physiotherapy room.

Physiotherapy Room 16 – 20m²

This room should allow for one or two therapists to work with individual pupils with physical difficulties.

Space will be needed for floor mats, large inflatables, parallel bars and other equipment, plus mirrors and desk space. If demand is low, some of these facilities can be provided in a large *SEN* resource base or a small group room used for other purposes.

PERSONAL CARE

See BB77 'Designing for Pupils with Special Educational Needs: Special Schools' 4.33.

Just as with furniture, there is no fit-all ergonomic design for toilets, wash basins, or showers. Different pupils will have different needs. Some might require one or two assistants to help them. Some might transfer laterally by themselves from a wheelchair to the toilet, requiring grab rails, back supports and other fittings. Others may just require grab rails for balance and clear signs on taps, towels and other equipment.

Toilets

- Toilets should be located so that distances pupils have to travel are not too great and routes are accessible.
- Toilets should be clearly signed.
- At least one toilet should be large enough to accommodate electric wheelchairs, assistants, and necessary equipment such as hoists and specialist fittings such as grab rails.
- There is detailed design guidance for toilets, which should be carefully followed and tested.
- The colour, or tone, of the background, fittings and any aids, such as grab rails, should be contrasted.
- Ceramic tiling and shiny floors may cause reflections and glare which might be confusing.
- The door of any toilet compartment should have the capacity to be opened outwards to ensure that entry can be gained even in event of someone falling and blocking the doorway.
- The lock mechanism, and whether there is a lock on the door, needs to be thought about in the light of pupils' physical design needs and cognitive abilities.
- Floors should be slip-resistant.
- Alarm systems should be installed at different levels,

including floor level, so pupils can call for assistance.

- A method should be established through the location of equipment and for staff for responding to any call for assistance from a pupil using a toilet.

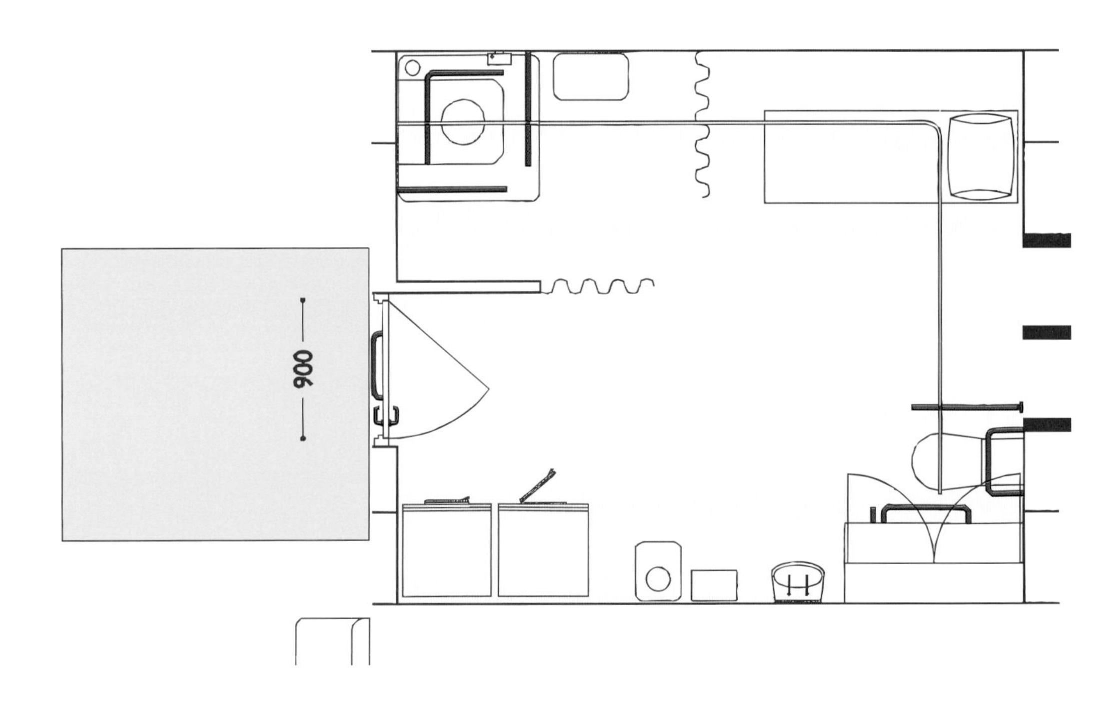

A hygiene room.

Hygiene Room 15 – 30m^2

This is necessary where the school accommodates pupils with severe physical difficulties or with profound and multiple learning difficulties. It should contain a shower, sluice, basic toilet, store cupboards, a changing trolley and room for assistants. A fixed or mobile hoist will also be required.

Laundry 5 – 8m^2

In schools where some pupils are incontinent, a small laundry will be essential. It must be separate from food areas.

See BB77 'Designing for Pupils with Special Educational Needs: Special Schools' 4.20 - 4.23.

TOILETS AND PERSONAL CARE PROVISION

See BB77 'Designing for Pupils with Special Educational Needs: Special Schools' 4.31.

Walney Secondary School caters for pupils between 11 and 16 including pupils with a wide range of *special educational needs* and *disabilities*. Facilities, such as resting rooms and teaching equipment, are

Building Bulletin 91 'Access for Disabled People to School Buildings' and Building Bulletin 77 'Designing for Pupils with Special Educational Needs: Special Schools' both provide guidance on toilets and personal care.

shared as well as skills and experience. In terms of the toilets and personal care provision, there are several important design features to note. The toilet is a large room with space for up to two adults and a pupil to manoeuvre. The seat is padded. A closomat is provided with bidet as well as hand drying equipment and a changing bed. An electrically operated ceiling hoist allows pupils to transfer themselves from the bed to the toilet.

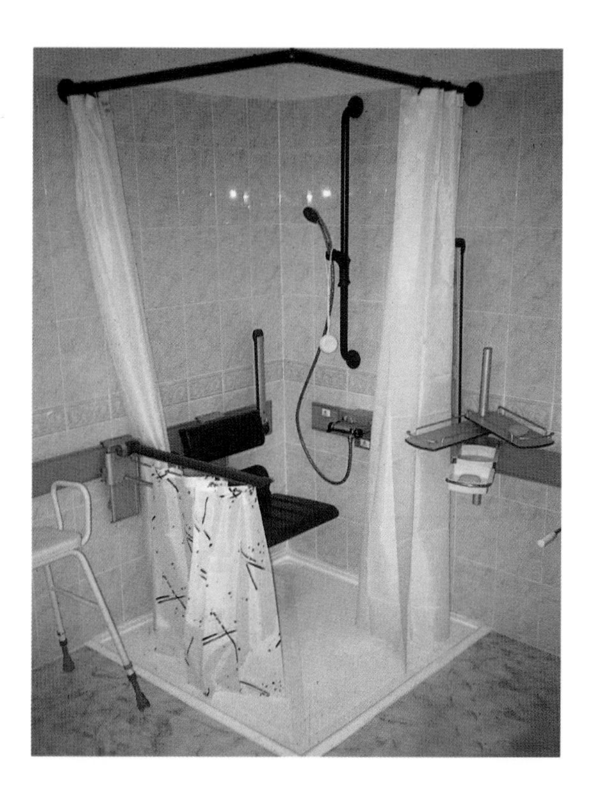

Accessible shower.

As with toilets there is detailed design guidance on creating accessible showers that should be followed. Any shower should be level, with no lip or rim, incorporate a seat at transfer level, a handset shower with lever operation, and thermostatic control with water temperature not exceeding 43°C.

OUTDOOR LANDSCAPES

With some people there is a growing interest in the creation of stimulating and varied school landscapes. Areas such as outdoor classrooms and sensory gardens are being included in school grounds alongside carefully designed seating areas, paths and planting.

- It is important to create a variety of spaces, including areas for small group discussions and quiet reflection.
- Boundaries need to be created between different activities, especially between higher and lower energy activities. These need not always be solid boundaries: seating, low planting and changes in surfaces can all give messages about what pupils can do in an area.
- As with indoor areas, consideration should be given to the nature of outdoor surfaces in terms of safety, colour, grip, texture and other characteristics.

Two types of playground, allowing for different levels of activity.

- Toxic plants should be avoided and spiky plants arranged so that pupils can easily identify them.
- Ways of maximising the sensory experiences of pupils should be explored. Scented plants, wind chimes, textured walls and seating, grass areas, robust and accessible furniture all could be included within the school grounds.

PLAY AND SPORT

Play is intrinsic to learning. To enable pupils to participate in the full range of play and leisure, outdoor games and sporting activities, it is important to consider:

Building Bulletin 71 'The Outdoor Classroom', Building Bulletin 85 'School Grounds: a guide to good practice' and publications from Learning through Landscapes all offer advice on inclusive landscape design.

- accessible changing rooms should include adequate space for assistants to help pupils change and shower in privacy;
- the ways in which play and sport areas can be made more accessible for pupils will vary according to their needs and aspirations.

INCLUSIVE SPORTS FACILITIES

Ormesby Secondary School has 800 pupils aged between 11 and 16. It accommodates pupils with a wide range of *special educational needs* and *disabilities*, including pupils with physical *disabilities*, *visual impairments*, speech and language difficulties and mild and severe *learning difficulties*. All pupils are fully integrated into the school.

The school has developed a wide range of facilities, including therapy suites and rooms where pupils can rest during the day. It has also developed inclusive sports facilities, which provide both inspiring examples of good practice as well as unanticipated problems to be tackled.

One of the most innovative aspects of the sports facility is the creation of adaptable changing areas. The two main changing areas sit alongside two smaller areas. These areas are not designated as male and female but used according to the groups of people using the centre and their particular needs. A second innovative design is the placement of the aerobics room next to an office, toilet, and store, which means it can be rented out to the health authority for use by mother and toddler groups.

The school found that a separate toilet compartment did not allow enough space for assistants, changing or for pupils to transfer themselves. Instead, a larger room equipped with a couch, hoist, sluice and toilet has been designed. Privacy when changing is important but so too is access. The school has found that using curtains rather than doors in changing areas whenever possible reduces problems of blocked access. Once a pupil has gone into a changing room, there needs to be a way they can communicate to an adult if they need help. The school is exploring two-way communication links.

Colour contrasting and good signage have been introduced to help pupils find their way and wherever possible, lighting is adjustable. Heights of furniture and equipment, such as basket ball nets and lockers, are also adjustable wherever possible. The centre is equipped with a climbing wall suitable for use by pupils with physical disabilities.

Swimming pool.

PLAY SPACES

Wigton Infant School caters for pupils between the ages of 4 and 7 with a wide range of difficulties, including *autistic spectrum disorder, hearing impairments,* severe *learning difficulties, physical disabilities* and *emotional and behavioural difficulties*.

The provision for indoor and outdoor play has improved in a number of ways since the school sought to address the needs of pupils with *special educational needs* and *disabilities*. Screens have been introduced to create enclosed spaces within classrooms where children can play alone or in small groups. A physical play room has been created which is shared by visiting therapists and pupils in the reception class. Similarly, a soft playroom is enjoyed both by pupils with *special educational needs* and *disabilities* and younger pupils in the reception class.

In terms of the outdoor play area, the school has examined the division of space, creating boundaries where needed between higher and lower energy play spaces.

Careful time-tabling of play time helps to avoid conflict between older and younger children. The school has also developed a very wide range of activities pupils can pursue outside the school, including riding, hydrotherapy and swimming.

PART 4 Useful Information

This final section provides additional, useful information.

4.1 PUBLICATIONS

PUBLICATIONS PRODUCED BY THE DEPARTMENT FOR EDUCATION AND EMPLOYMENT

Circular 20/99 What the Disability Discrimination Act 1995 means for Schools and Local Education Authorities.

SEN Code of Practice on the Identification and Assessment of Pupils with Special Educational Needs and SEN Thresholds.

Good Practice Guidance on Identification and Provision for Pupils with Special Educational Needs, consultation document, July 2000.

ISBN 0-11-271015-8

The SENCO Guide, 1997.

ISBN 0-11-271061-1

Building Bulletin 71:
The Outdoor Classroom, Second Edition. TSO 1999

ISBN 0-11-270796-3

Building Bulletin 77:
Designing for Pupils with Special Educational Needs: Special Schools. HMSO 1992.

ISBN 0-11-270921-4

Building Bulletin 82:
Area Guidelines for Schools. HMSO 1996.

ISBN 0-11-270990-7

Building Bulletin 85:
School Grounds: a guide to good practice. TSO 1997.

Building Bulletin 87:
Guidelines for Environmental Design in Schools.
TSO 1997.
ISBN 0-11-271013-1

Managing School Facilities
Guide 7: Furniture and Equipment in Schools:
a purchasing guide. TSO 2000.
ISBN 0-11-271092-1

Building Bulletin 90:
Lighting Design for Schools. TSO 1999.
ISBN 0-11-271041-7

Building Bulletin 91:
Access for Disabled People to School Buildings.
TSO 1999.
ISBN 0-11-271062-X

Designing for 3 to 4 year olds: guidance on
accommodation for various settings. TSO 1999.
ISBN 0-85522-986-1

OTHER PUBLICATIONS

Best Play, The Children's Play Council with
The National Playing Field Association and
Playlink, 2000.

Index for Inclusion: developing learning and
participation in schools, Centre for Studies in Inclusive
Education in collaboration with the Centre for
Educational Needs.

Children with Disabilities.
The Disability Information Trust (DIT) has produced a
new publication entitled 'Children with Disabilities'.
This 200 page, illustrated book provides information
on the wide range of equipment designed to enable
children with disabilities to perform everyday tasks that
others take for granted. The publication costs £11 and
is available from the DIT:

Mary Marlborough Centre
Nuffield Orthapaedic Centre
Headington
OXFORD OX3 7LD
Tel: 01865 227 592
Fax: 01865 227 596
E-mail: ditrust@btconnect.com

Grounds for Sharing: a guide to developing special school sites, Learning through Landscapes, 1996. This publication costs £13.50 plus £1.25 postage and packing and is available from:
Learning through Landscapes
Third Floor
Southside Offices
The Law Courts
WINCHESTER
Hampshire SO23 9DL

National Curriculum: Handbook for Primary and Secondary Teachers in England. Free to schools from DfEE Publications Centre Tel: 0845 6022260

4.2 CONTACTS

The Special Educational Needs and Disability Bill. The proposed legislation on special educational needs and disability rights in education will make it unlawful for education providers to discriminate against a disabled child by:

- treating a disabled child less favourably because of his/her disability in the arrangements made for the provision of their education;
- failing to take reasonable steps to change any policies, practices or procedures which place a disabled child at a substantial disadvantage compared to a non-disabled child;
- failing to take reasonable steps to provide education using a reasonable alternative method where a physical feature places a disabled child at a substantial disadvantage compared to a non-disabled child.

For information and advice concerning all aspects of the implementation of the Disability Discrimination Act, contact the Helpline.

DDA Help
FREEPOST
MIDO 2164
STRATFORD-UPON-AVON CV37 9BR
Live Operator: 0345 622 633
Text phone: 0345 622 644
Faxback: 0345 622 611
Automated: 0345 622 688
E-mail: ddahelp@stra.sitel.co.uk

All DDA-related publications, including the Code of Practice: Rights of Access, Goods, Facilities, Services and Premises can be downloaded from the website: www.disability.gov.uk

Aiding Communication in Education (ACE) Centre Advisory Trust. Provides a focus for the use of technology with the communication and educational needs of young people with physical and communication difficulties.
92 Windmill Road
Headington
OXFORD OX3 7DR
Tel: 01865 759800
Fax 01865 759810
E-mail: info@ace-centre.org.uk

Architects and Building Branch, DfEE,
Caxton House
6-12 Tothill Street
LONDON SW1H 9NA
Fax: 020 7273 6762
Enquires: Nicola Williamson Tel: 020 7273 6023
E-mail: nicola.williamson@dfee.gov.uk
Website: www.dfee.gov.uk/schbldgs

Association of Workers for Children with Emotional and Behavioural Difficulties (AWCEBD)
Allan Rimmer
Administrative Officer
Charlton Court
East Sutton
MAIDSTONE ME17 3DQ

Tel: 01622 843104
Fax: 01622 844220
Website: www.mistral.co.uk/awcebd

CENMAC The Centre for Micro-Assisted Communication. Provides support and advice on using Information Technology to improve the communication and learning of children and young people with physical disabilities.
Charlton Park School
Charlton Park Road
LONDON SE7 8JB
Tel: 020 78854 1019
Website: www. cenmac.demon.co.uk

The Center for Universal Design.
Universal Design creates products and environments usable by all people, to the greatest extent possible, without the need for adaptation or specialised design. Provides national (US) research, information and technical assistance that evaluates, promotes and develops universal design in housing, public and commercial facilities.
E-mail: cud@ncsu.edu
Website: www.design.ncsu.edu

Centre for Accessible Environments.
Gives advice and information on making buildings and places accessible to all users and on enhancing quality in design. The Centre provides access consultancy and produces technical design guides and other publications.
Nutmeg House
60 Gainsford Street
LONDON SE1 2NY
Tel: 020 7357 8182
E-mail: cae@globalnet.co.uk
Website: www.cae.org.uk

Centre for Studies on Inclusive Education.
British independent educational charity gives advice and information about inclusion and related issues. Provides a useful leaflet entitled Money for Inclusion

that describes the workings of the Standards Fund, the Schools Access Initiative and Out of Hours Activities Funding.
1 Redland Close
Elm Lane
Redland
BRISTOL
Avon BS6 6UE
Tel: 0117 923 8450
Fax: 0117 923 8460
Website: www.inclusion.uwe.ac.uk

Children's Play Council.
A specialist organisation under the National Children's Bureau. Promotes importance of consultation with children and young people of all abilities, and it stimulates partnerships between play and other services to children and young people.
Tel: 020 7843 6016
E-mail: cpc@ncb.org.uk

Council for Disabled Children.
Works to empower children with special needs and their families. Promotes collaborative work between different organisations providing services and support for children and young people with disabilities and special educational needs.
c/o National Childrens Bureau
8 Wakey Street
LONDON EC1V 7QE
Tel: 020 7843 6061
E-mail: cdc@ncb.org.uk.

The Disabled Living Foundation.
Provides fact sheets on choosing grab rails, stair lifts, hoists, children's play and mobility equipment.
380-384 Harrow Road
LONDON W92HV
Helpline: 0870 6039 177
Website: www.dlf.org.uk
E-mail: dlfinfo@dlf.org.uk

The Dyspraxia Foundation
8 West Alley
HITCHIN
Hertfordshire SG5 1EG
Helpline: 01462 454 986
Tel: 01462 455 052

Learning through Landscapes.
Dedicated to helping schools improve their grounds for the benefit of all children.
3rd Floor, Southside Offices
The Law Courts
WINCHESTER SO23 9DL
Tel: 01962 846 258
Fax: 01962 869 099
Website: www.ltl.org.uk

Makeover at School (M@S).
Inter-disciplinary organisation of SEN experts, architects and educational researchers working to explore the relationship between school buildings, learning and behaviour. Carrying out research based and practical design projects, M@S promotes the idea that school buildings that contribute to education in its widest sense are an essential element of inclusion.
Helen Clark, Project Co-ordinator
SENJIT
Institute of Education
University of London
20 Bedford Way
LONDON WC1H OAL
Tel: 020 7612 6273/4
Fax: 020 7612 6994
E-mail: h.clark@ioe.ac.uk

Mencap.
Works with people with learning disabilities to fight discrimination.
123 Golden Lane
LONDON EC1Y 0RT
Tel: 020 7454 0454 / 020 7696 5503
Fax: 020 7696 5540
Website: www.mencap.org.uk

National Advisory Group On SEN (NAGSEN).
Set up in July 1997 to take forward the Government's commitments on raising the standard of education for children with SEN
Website: www.dfee.gov.uk/sen/nagsen

National Deaf Children's Society (NDCS).
15 Dufferin Street
LONDON EC1Y 8UR
Tel: 020 7250 0123
Fax: 020 7251 5020

Network 81.
An umbrella organisation of support groups to educate parents about all aspects of the education of their children with special needs.
Helpline Monday to Friday 10.00am to 2.00pm
Tel: 01279 647415
Fax: 01279 814908
E-mail: network81@tesco.net

Qualifications and Curriculum Authority (QCA).
The government agency responsible for advising the Secretary of State on matters to do with curriculum, assesements and qualifications.
83 Piccadilly
LONDON W1J 8QA
Tel: 020 7509 5555
Fax: 020 7509 6666

RADAR.
Provides information on the needs of disabled people and consultancy services.
12 City Forum
250 City Road
LONDON EC1V 8AF
Tel: 020 7250 3222
Fax: 020 7250 0212
E-mail: radar@radar.org.uk
Website: www.radar.org.uk

Royal Institute of British Architects (RIBA) Schools' Client Forum.
Website: www.riba.net
Then search for Schools' Client Forum and go to publications page.

Royal National Institute for the Blind (RNIB).
224 Great Portland Street
LONDON W1N 6AA
Tel: 020 7388 1266 (switchboard/overseas callers)
0845 766 99 99 (UK Helpline)
Fax: 020 7388 2034
website: www.rnib.org.uk

Royal National Institute for Deaf People (RNID).
Provides consultancy on the environmental needs of people with hearing impairments.
19-23 Featherstone Street
LONDON EC1Y 8SL
Tel: 0870 605 0123
Fax: 020 7296 8199
helpline@rnid.org.uk
website: www.rnid.org.uk

SCOPE.
Educational services provide training and support for children, their parents and helpers and adults with cerebral palsy.
6 Market Road
LONDON N7 9PW
Tel: 020 7619 7100
website: www.scope.org.uk

SENSE.
Leading organisation working with people who have sight and hearing difficulties.
11-13 Clifton Terrace
Finsbury Park
LONDON N4 3SR
Tel: 020 7272 9648
Fax: 020 7272 9648
E-mail: enquiries@sense.org.uk

Special Educational Needs Division, DfEE.
Part of the National Grid for Learning (NGfL)
Website: www.dfee.gov.uk/sen

4.3 RESEARCH PROCESS

The project received valuable support and input from the Steering Group. Members of the Steering Group included:

Cathie Bull, HMI, OFSTED
Helen Clark (Project Co-ordinator), SENJIT
Dominic Cullinan (Consultant), CABAL
Mike Diaper, SEN Division, DfEE
Flora Gathorn-Hardy (Consultant)
David Hayhow (Consultant), David Hayhow Architects
Mary Hrekow (Consultant), SENJIT Co-ordinator
Colin Jefferson, Property Officer, Cumbria LEA
Mukund Patel, Head of Architects and
Building Branch, DfEE
Nick Peacey, Equal Opportunities Manager, QCA
Caroline Roaf, NASEN
Philippa Russell, Council for Disabled Children,
National Children's Bureau
Robin Thomas, SEN Manager, Hampshire
Peter Weston, Assistant Director, Education
Department, East Sussex
Beech Williamson, Architects and
Building Branch, DfEE

Research visits were made to the following schools.

Alexandra Special School, Harrow
Binfield Primary School, Hampshire
Bodmin Community College, Cornwall
Filsham Valley School, East Sussex
New Tulse Hill, Lambeth
Ormesby Comprehensive School, Middlesbrough
Pennyman Primary School, Middlesbrough
Treviglas Community College, Cornwall
Ullswater Community College, Cumbria
Vickerstown Primary School, Cumbria

Walney Secondary School, Cumbria
Westgate Community School, Hampshire
Wigton Infant School, Cumbria
West St Leonard's Community Primary School, East Sussex
Whitmore High School, Harrow

Representatives from the following Local Education Authorities were also consulted:

Former Avon
Paul Gillespie, Team Leader, Hearing Impairment team
Sue Rogers, Team Leader, Visual Impairment team
Sally Silverman, Multi-Needs Sensory Impairment team

Birmingham Advisory and Support Services
Judith Gibbons, Disability Access Co-ordinator
Joyce Nicholls, Chair of Access Group

Cornwall Education
Gerald Adye, Head of Schools Section
Ken Allen, Schools Surveyor
Geoff Hogg, Head of Individual Needs

Gloucestershire
Mark Geraghty, Head of Sensory and Language Impaired Service.

Hampshire
Alec Gillies, Architect

Hertfordshire
Honor Anderson, SEN Advisor

OTHER ORGANISATIONS CONSULTED

Helen Allen, Access Consultant, Joint Mobility Unit
Richard Vaughan, Technical Advisor, National Deaf Children's Society
Mark Vaughan, Centre for Studies on Inclusive Education
Ken Black, Inclusive Sport Officer, Youth Sports Trust

4.4 GLOSSARY

Acuity
Clarity or sharpness of vision.

Autistic spectrum disorder (ASD)
A condition which is characterised by: the absence or impairment of two-way social interaction; absence or impairment of comprehension and the use of language and non-verbal communication and lack of imagination/abstract thought. Symptoms may include obsessive behaviour, self-injury, attachment to objects etc.

Co-location
A term to describe the sharing of a site and facilities by a *special school* and a *mainstream school*.

Disability
The Disability Discrimination Act 1995 defines a disabled person as someone who has 'a physical or mental impairment which has a substantial and long-term adverse effect on his ability to carry out normal day-to-day activities'. This is the definition which will apply to pupils and students when the proposed disability discrimination in education legislation comes into force.

Dyslexia
A specific learning difficulty relating to literacy.

Dyspraxia
A developmental co-ordination disorder, symptoms of which include clumsiness, poor posture, poor short term memory and reading and writing difficulties.

Emotional and behavioural difficulty (EBD)
Emotional and behavioural difficulties are *special educational needs*, which include pupils who are excessively quiet and withdrawn and those whose behaviour is consistently inappropriate. Emotional and behavioural difficulties lie on a continuum between behaviour which is disturbing and challenges teachers but is within normal bounds and that which is indicative of serious mental illness. Risk factors

for EBD include other causes of learning difficulty and adverse family and environmental factors.

Hearing Impairment (HI)
The loss of ability to hear normally - severity varies from slight, to moderate, to severe and finally profound. Hearing loss may be 'sensori neural' i.e. due to absent or damaged auditory nerves, or 'conductive' due to a disorder of the middle ear. The latter is much more common but is less severe.

Inclusive Learning Environment
A learning environment designed to provide access to the curriculum for all pupils.

Induction loops
Electronic loop set on the walls or ceilings of a teaching or other space which enables users of hearing aids to pick up signals.

Infrared system
A wireless sound transmission system which conveys sound from its source to hearing aids

Learning difficulty
Pupils with learning difficulties are those whose general level of attainment and rate of progress will be significantly below that of their peers.

Mainstream schools
A mainstream schools is defined by section 316 of the Education Act 1996 as a 'school which is not a special school' and where the Local Education Authority has ultimate responsibility to carry out the functions undertaken by the governing body in a maintained primary or secondary school.

Mobility aids
Aids which enable visually impaired pupils to move safely and independently.

Multi-Sensory Impairment
Also known as 'deaf blind'.

National Curriculum
Statutory defined elements of the school curriculum.

Radio Systems
Form of sound enhancement in the classroom for particular pupils. The teacher has a microphone and the pupil has a receiver.

Resource Base
A specialist resource base, which may be attached to a mainstream or special school. A resource base might provide access to specialist education and health staff, specialist equipment and resources to enable access to the curriculum. A resource base can also provide specialist advice, support and training for school staff on appropriate curriculum planning and effective teaching and learning for pupils with specific needs. Some resource bases will also provide specialist small group or individual teaching and/or therapy.

School development plan
A school's plan for raising educational achievement through setting out the priorities and timing of the school's schedule, usually over several years.

SENCO
Special educational needs co-ordinator. A teacher with responsibility within a school for: the day-to-day operation of the school's SEN policy; liaison with, and advice to, other staff; co-ordinating provision; maintaining the school SEN register and overseeing SEN records; liaison with parents and external agencies.

Sensory impairment
A hearing or visual impairment or combination of the two.

Sensory room
Environments structured to facilitate multi-sensory experiences through the use of light, tactile modifications and auditory access.

Sign and symbol system
Sign - language using the hands for communication.
Symbol - use of symbols to represent words or concepts.

Sound field amplification systems
A form of enhancing sound in the classroom usually using a teacher with a microphone and a grid of speakers set into the classroom ceiling.

Special educational needs (SEN)
Pupils have *special educational needs* if they have learning difficulties which calls for special educational provision to be made for them. This may be because they have significantly greater difficulty learning than the majority of pupils the same age. The full definition is in section 312 of the Education Act 1996.

The Special Educational Needs Code of Practice
The 'Code of Practice on the Identification and Assessment of Special Educational Needs', known as the 'SEN Code of Practice', is the guidance produced under the provisions of Section 313 of the Education Act 1996 (previously part of the 1993 Education Act). The Code gives practical guidance to local education authorities and the governing bodies of all maintained schools on their responsibilities towards all children with SEN. This includes guidance on policies and procedures aimed at enabling pupils with *special educational needs* and *disabilities* to reach their full potential, to be included fully in their school communities and make a successful transition to adulthood.

Visual Impairment
Any disorder which affects the eyesight ranging from partial sight to blindness.

Published by The Stationery Office and available from:

The Stationery Office
(mail,telephone and fax orders only)
PO Box 29, Norwich, NR3 1GN
Telephone orders/General enquiries 0870 600 5522
Fax orders 0870 600 5533

www.thestationeryoffice.com

The Stationery Office Bookshops
123 Kingsway, London WC2B 6PQ
020 7242 6393 Fax 020 7242 6394
68-69 Bull Street, Birmingham B4 6AD
0121 236 9696 Fax 0121 236 9699
33 Wine Street, Bristol BS1 2QB
0117 926 4306 Fax 0117 929 4515
9-21 Princess Street, Manchester M60 8AS
0161 834 7201 Fax 0161 833 0634
16 Arthur Street, Belfast BT1 4GD
028 9023 8451 Fax 028 9023 5401
The Stationery Office Oriel Bookshop
18-19 High Street, Cardiff CF1 2BZ
029 2039 5548 Fax 029 2038 4347
71 Lothian Road, Edinburgh EH3 9AZ
0870 606 5566 Fax 0870 606 5588

The Stationery Office's Accredited Agents
(see Yellow Pages)

and through good booksellers